Letters from the Future

Letters from the Future

How New Brunswickers Confronted Climate Change and Redefined Progress

Edited by
Daniel Tubb
Abram Lutes
Susan O'Donnell

Illustrated by Ian Smith

Chapel Street Editions

Copyright © 2021 by the authors
Illustrations copyright © 2021 by Ian Smith

Published by
Chapel Street Editions
150 Chapel Street
Woodstock, NB E7M 1H4
www.chapelstreeteditions.com
chapelstreeteditions@gmail.com

ISBN: 978-1-988299-37-2

Library and Archives Canada Cataloguing in Publication

Title: Letters from the future : how New Brunswickers confronted climate change and redefined progress / edited by Daniel Tubb, Abram Lutes, Susan O'Donnell ; illustrated by Ian Smith.
Other titles: Letters from the future (2021)
Names: Tubb, Daniel, editor. | Lutes, Abram, editor. | O'Donnell, Susan (Editor), editor. | Smith, Ian (Ian Robert Laird), 1956- illustrator.
Description: Includes bibliographical references.
Identifiers: Canadiana 20210307188 | ISBN 9781988299372 (softcover)
Subjects: LCSH: Sustainable living—New Brunswick. | LCSH: Climate change mitigation—New Brunswick.
Classification: LCC GE196 .L48 2021 | DDC 333.7209715/1—dc23

Book Design by Brendan Helmuth

Illustrations by Ian Smith

Chapel Street Editions, Ltd. gratefully acknowledges the financial support of the Department of Tourism, Heritage, and Culture, Province of New Brunswick.

Dedication

For all New Brunswickers...
 ...now and in the future.

Contents

Preface i
Acknowledgements iii

Letters from the Future 1

Part One: It Begins with Land and Water 3

    Dear Grandmothers
        Terry-Ann Sappier 5

    The Tale of Skutik
        The Skutik Collective 7

    Ocean Assist Dive Log
        Renelle LeBlanc 11

    Twenty Shades of Green
        Tom Beckley 13

    Flood Memories and Graveyard Parks
        Mary Louise McCarthy-Brandt 16

Part Two: Shaking the Institutions 21

    What was their Real Name Again?
        Alain Deneault 23

    Truth-fullness
        Ajay Parasram 25

    Who Saved the World? Girls!
        Lauren R. Korn 29

    Electoral Reform
        Victoria Clowater 32

    The New Brunswick Youth Parliament
        Arianne Melara Orellana .35

    The Revolution of a New Generation
        Alicia F. Noreiga 38

Part Three: Welcoming Newcomers and Coming Home 43

    They Came in Waves
        Jael Duarte. 45

    Go East, Young Woman
        Leland Wong-Daugherty 48

    How Far We've Come!
        Sagda Elnihum 51

    What's Changed? A Lot!
        Sara Taher 53

    A Letter from El Salvador
        Carlos Morales 55

    We Came, We Saw, We Helped Build the Economy
        Lisa-Gay Taylor. 57

Part Four: Building New Foundations .61

    A Guaranteed Annual Income
        Amy Floyd. 63

    Rural Assemblies
        Abram Lutes 66

    Organizing Food Security
        Louise Livingstone 70

    Ending Homelessness
        Nigam Khanal. 72

    Transforming the Nature of Work
        Christine Wu 74

    Education First: Everything Else Follows
        Naomi Gullison. 77

    Restorative Justice: The Way Forward
        LA Henry 80

Part Five: A Resilient, Low-Carbon Way of Life 85

    New Brunswick's Green New Deal
        Daniel Tubb 87

    And then Faster than We Thought Possible
        Erin Seatter 90

    Watershed Councils Adapting to the Climate
        Adje Prado. 93

    The Flip to Low-Carbon Mobility
        Matthew Hayes 96

    Continuing the Slow Decline, but that's Okay
        Cheryl Johnson 99

    Two Visions of the Future
        Carl Duivenvoorden 102

    Connected Communities
        Susan O'Donnell 104

Part Six: Strong Relationships Make Dreams Possible 109

    Green-Shifting Education
        Raissa Marks 111

    Taking Back our Resources
        Chris Rouse 114

    The Joys of Rural Work
        Teri McMackin 117

    Local Food, for Local Communities
        Stephanie Coburn. 120

    Gardening the Margins
        Kylie Bergfalk. 122

    Blessings from 2066
        Jean Desrosiers 124

Afterword: Change the Story, Redefine Progress, Create the Future 127

About the Authors, Editors, & Illustrator 135

Preface

New Brunswickers are hungry for hopeful stories of the future to counter the stereotype of "hard times in the Maritimes." With the climate changing and a pandemic transforming life as we knew it, the future now seems very uncertain. But now is the time to imagine a better future. The future is created from what we imagine is possible. With the right stories to guide us, we can create a healthier, happier province, richer in the ways that matter. *Letters from the Future* presents a sample of these stories, written by people who care deeply about New Brunswick

The stories detail the hopes and aspirations of New Brunswickers from different backgrounds and different parts of the province. The hopes are for a more equitable New Brunswick, a home where sustainable farming and fishing flourish, where food is secure, where our rural communities, towns and cities are vibrant, where natural resources are used wisely and sustainably, where governance is democratic, and where public projects are community-focused, innovative, and change life for the better. The future our authors envision feature sustainable livelihoods in thriving communities.

Our authors transported themselves to the distant and not-so-distant future and describe what it took to get us there. Our letters dream about what New Brunswick could become if we seize the opportunity to create the province we want to live in — a New Brunswick that confronts climate change and achieves its goal of sustainable living, and a province that helps effectively address the environmental, economic, social justice, human rights, and public health challenges we are currently experiencing.

Each letter offers a different aspect of a possible future. The power of speculative nonfiction reveals what is possible, showing alternatives not yet realized. Dystopic nightmares need not come to pass — although several authors write about such conditions. The fact that New Brunswickers are now beginning some of the great changes envisioned in these letters illustrate that our creative speculations are based on what is achievable in the future.

The changes required for realizing the futures imagined in this book are profound. The COVID-19 pandemic, the changing climate and extreme weather events, our unsustainable economy, the growing and extreme disparities between rich and poor, increasing homelessness and ongoing social and economic marginalization show us that the existing systems are not working for most of us. What better time than now to imagine and build an alternative path into the future?

Our letters employ different systems of thought to solve complex challenges. Rather than proposing silver bullets, the letters offer diverse and sometimes contradictory narratives of green futures adapted to local conditions. Each letter is its own story. Each looks back to imagine what New Brunswick could accomplish if we change the narrative, fight for the future, and win. Each letter applies progressive ideas from the global zeitgeist to local problems.

The letters in our book add a unique collective voice missing from the ongoing discussion in our province about who we are and where we should be going. Our stories push back against the blinkered visions offered by many politicians, pundits, corporate voices, and think-tanks, while overcoming the cynicism common among critics of the status quo.

Our letters are the opposite of cynical. We offer hope. We propose inspiring visions that challenge low expectations and worst-case scenarios for New Brunswick. We dare to assert that our narratives about the future not only matter but also build a path to a better New Brunswick for us all. We hope this book finds a home on the shelves of New Brunswickers and people everywhere who are hungry for alternatives.

Daniel Tubb
Abram Lutes
Susan O'Donnell

Acknowledgements

Anthropologist Margaret Mead once observed: "Never doubt that a small group of thoughtful, committed, citizens can change the world. Indeed, it is the only thing that ever has." The editors of *Letters from the Future* are such a group. We, and all the authors, are also members of many other small, committed groups embedded in the New Brunswick communities in which we live.

The letters that compose this book emerged from a collaboration between the three editors as part of the research project Rural Action and Voices for the Environment (RAVEN) at the University of New Brunswick, in partnership with the New Brunswick Media Co-op. The NB Media Co-op is an independent, not-for-profit, online and print news publication run by volunteers and celebrating twelve years of continuous operation. The NB Media Co-op relies on memberships and individual donations as well as support from community partners. RAVEN is funded by the Social Sciences and Humanities Research Council of Canada and the New Brunswick Innovation Foundation. The editors are all members of the RAVEN research team and active with the NB Media Co-op.

The idea for *Letters from the Future* was born in the spring of 2019 after the 2018 landmark report from the Intergovernmental Panel on Climate Change and in the months before a federal election. The IPCC report gave us until 2030 to make the required changes for avoiding climate catastrophe.

We invited friends, colleagues, and others to write letters from the future and began publishing the series online with the NB Media Co-op. About half the letters in this book were published online in 2019, the remainder solicited in 2020. The authors encompass a broad demographic: farmers and small business owners, people living in off-grid communities, newcomers to the province, members of First Nations, New Brunswickers of African descent, university professors, students, activists, and politicians.

More than half the letters in this book were written before the province shut down because of the COVID-19 pandemic. While this makes some letters seem out of date, or naïve, it makes others seem to come true overnight; for example, Susan O'Donnell's letter, which included the closure of university campuses in favour of online education.

To all contributing authors, we offer our gratitude. We hope you found the process of writing your letter inspiring and rewarding. We also hope that re-reading your letter and reading those of other authors will be as inspiring and exciting as it has been for us to imagine a positive and flourishing future for New Brunswick.

As letters were arriving and the structure of the book was taking shape, we commissioned Ian Smith, a New Brunswick artist and outdoor educator, to add a complementary vision to the letters, to tell a graphic version of the story evolving throughout the book. We are grateful to Ian for his enthusiastic contribution to this project.

The book would not have emerged without the encouragement of Tracy Glynn and her help in finding a publisher. The keen response of Keith Helmuth, Managing Editor at Chapel Street Editions, and the positive reception by all our authors for the publication of a book, inspired us to do the work of soliciting more letters, then editing, organizing, and finishing the project.

Letters from the Future has benefitted from the careful and deft editing of Karen Caruana, and the encouragement of many people who wanted to see this book come to fruition.

Thank you all.

Letters from the Future

Part One
It Begins with Land and Water

Part One: It Begins with Land and Water

Dear Grandmothers
Terry-Ann Sappier

I am writing to offer you my gratitude for my life and for those who will come after me. I am the seventh generation Granddaughter of one of the Grandmothers who defended this land. My parents and Grandparents made sure your story was passed down so we would never forget the sacrifice you made for our future. For that reason, I am writing this letter.

We were told of the courageous efforts made to ensure we had the same rights and responsibilities our Forefathers envisioned for us when they signed the sacred Treaties. Now we live as one people, all Treaty people, each understanding the obligations and responsibilities set forth.

Our Ancestral homeland is abundant and provides all we need to survive and live in a good way. We have food security, fresh clean drinking water, trees that your arms only go halfway around the trunk when you hug them. With the hydro dams gone, the salmon have returned to the rivers—so many we only need to grab them with our hands. But don't worry, we only take what we need.

We power our communities with local green energy, wind, solar, and geothermal. We have no carbon emissions. Our sky is the prettiest blue and filled with birds of all colours and sizes. Because of your vision for a better tomorrow, we can embrace clean living. Your unwavering need to ensure we had a future makes me most proud to be your descendant.

There is a place near Napadogan that is protected. We were told a camp was erected there to protect the headwaters and land from the tailings pond of a mining project. Had you not stopped the project from moving forward, we would be living with irreparable damage. Now the land is protected and flourishes with wildlife.

The fossil fuel age died along with the need to extract resources. We reuse and recycle everything—no waste. The age of consumerism has ended. We all have what we need to live a good life. Excessive accumulation is no longer practiced. Everything we need is in the community. We support one another and take care of one another, much like in your ancestor's time, the old days.

I know it was your love for this land and the people of this land that gave you the strength you needed to stand strong and true for me and all New Brunswickers to enjoy and live a clean healthy life. The sacrifices you made are inconceivable. We owe you a great debt. Without you our forests would be tree farms for industry, our waters poisoned with chemicals and dead to all life, and our farms sprayed with poisons that give people cancer. There are no more cancer zones in New Brunswick. Once the spraying stopped, people started living longer healthier lives. I wish you could see the future you have helped create. Because of your commitment and tenacity, you changed the way New Brunswickers think and live. You paved the way for a better tomorrow.

New Brunswick is now Canada's most visited province. Our main economy is tourism. People come from all directions to enjoy the natural beauty of our province. There are adventures for all seasons and all types of activities for people to enjoy. New Brunswick has become a role model for the other provinces and territories, going from a "have-not" province to a "have" province. As a leader in green technology and an innovator in the sustainable development of our lands, we have been able to move our province into economic prosperity.

With the true intent of the Treaties entrenched in our structure of government, the province is now co-managed, and everyone is prosperous. Finally, reconciliation has been fulfilled. We are one people, proud and determined to maintain the life we have all worked hard to bring together for the betterment of all life, not just humankind.

The population has grown substantially over the last one hundred years. People come to New Brunswick from all countries, escaping the pollution and crowded cities. Our urban centres are not as large as those in other provinces, no skyscrapers or tall buildings. We embrace rural living. We grow most of our food and, using new technology, we grow year-round. We do import some produce but not on the scale it was in your time.

So, you see my Dear Grandmothers, you gave us the future you all strived to give. Your efforts have not gone unnoticed or forgotten. We are so grateful for all that you did, stopping mining projects, fracking, hydro dams, and spraying, the list goes on. Thank you for all you did for us because you loved us.

Part One: It Begins with Land and Water

The Tale of Skutik
The Skutik Collective

I am the largest river between the Penobscot and Wolastoq (Saint John) watersheds and the heart of the Peskotomuhkatik Ancestral Homeland, an area of more than three million acres. I am known as, and will answer to, St. Croix or Schoodic, but my spirit is carried by my honoured name of Skutik, "the home of the big fire."

I descend from the Chiputneticook Lakes and flow 114 kilometres south to the Passamaquoddy Bay. The settlements along my shores were born of my assets and because of my ecological and navigational links to the Bay.

I am a river, I am an estuary, and I am life. I have been given the responsibility to act as an international border, an example of the two-leggeds' desire to organize and control. For centuries I was, and I am now again, noted for my large runs of anadromous fish migrating upriver from the sea to spawn. But it was not always so.

As you listen to my story—the story of my air and water clans, and my two-leggeds—hear from your heart, hear the drums ringing out along the valley as they did for years, hear the drums that represent my heartbeat, the rhythm of life.

Breaking with tradition, I start with the moral of the story—the connection to me. For all those within me, and to my terrestrial brother and sister clans, this is the foundation for all life, for all health, and for all economies. As my reserves were drained, so too were the reserves of strength and hope of my two-legged people. Damming me (literally) and damning me (figuratively) not only transformed my natural landscape but also ushered in an era of pollution, degradation, and neglect. Both changed not only me but also the attitudes of two-leggeds to each other. Their actions and attitudes were bred through generations of settlers, and the ability to thrive became out of reach for me and for all of us.

My landscape, the Skutik region, embodies the connection between a thriving environment and thriving communities. I understand these interdependencies well—the flows of respect and the ability for interdependent systems to thrive. I established myself here at the end of the last Ice Age. From this beginning long ago, let me tell my tale, the tale of Skutik.

Once I was tickled by the canoe bottoms of my people. My two-leggeds made arrangements with the fish who lived within me to collect them for their subsistence—the *psam* (shad), *siqonomeq* (alewife), *polam* (salmon), *kat* (eel), and *pasokos* (sturgeon). This changed when we welcomed "visitors" from away. When these "visitors" became permanent settlers, the strength of the Peskotomuhkati Nation was decimated by disease and by their isolation from the tree clans and the four-leggeds and from me.

The injuries were plenty and came in the form of dams that supported an era of industrialization starting in 1793 in my upper reaches. Before the early European settlement of this country until 1825, there was annually a great abundance of *psam*, *siqonomeq*, *polam*, *kat*, and *pasokos*. Boats from away, some more than 150 tons, took too many members of my water clans. The boats never left without full cargo holds.

When I thought I could bear no further insult, the settlers had more in store for me. In 1825 they built the Union Dam on my lowest reaches with no fishway, and my anadromous brothers and sisters of the water clan could visit me no more. The refuse from upriver sawmills created new islands within me. Only after this sawdust waste created navigational hazards, and long after my bottom-dwelling cousins had suffocated and died, did the two-leggeds pay attention. In addition to my water-borne family being desecrated, many of the land-dwelling tree clans were lost during the log drives and lay dishonoured on my bottom.

By the 1860s my waters were being poisoned with human waste, along with salt liquids, lime liquor, and skin scrapings from tanneries upriver. By 1934, I could take it no more. I summoned all my reserves and called in the wind clan as well. We breached the Union Dam. Only three dams were left standing on my lower reaches, and I could hear the people say the names *psam, siqonomeq, polam, kat,* and *pasokos*. The call for my rehabilitation commenced.

However, conditions worsened through to the 1960s. Over many decades, the two-leggeds could not make up their minds. During this time, I had intermittent fishways put in and the two-leggeds stocked me with more than a million salmon from Wolastoq waters and other places, but never did they clean up the sawdust, or the chemicals, or their behaviour. My recovery was slow. The two-leggeds found mill sludge deposits seven feet deep on my bottom. In the fishways my water children perished within an hour due to lack of dissolved oxygen; they could not breathe.

By 1975, my toxic exhalations were peeling paint off nearby houses and damaging the lungs of the two-leggeds living along my shores. But those defending my life stood firm and never gave up. They believed I could be who I am now. They carried on with plans to restore my health and restore the access of the water clans to my watershed. But I had to face one last fight.

In the 1980s, the smallmouth bass introduced in my home lakes by the two-leggeds were in substantial decline. Local guides and sporting camps who had developed a livelihood based on these newcomers complained and lobbied the American government. Fishways were closed to the *siqonomeq* who stood accused for the decline of smallmouth bass. There was a complete decimation of my *siqonomeq* run to less than 1,000 fish in 2002. In Canada, the Department of Fisheries and Oceans trucked my *siqonomeq* children from the Milltown Fishway to the Woodland impoundment in an attempt to save them.

I thought I had no more fight left in me, but my allies rallied and took me with them.

In 1991, I received a designation as a Canadian Heritage River. I was recognized for my outstanding natural, cultural, and recreational heritage, and for my important habitats that support rare plants and freshwater and marine invertebrates, amphibians, reptiles, and birds.

My River Keeper brethren, the Peskotomuhkati Nation, and others fought hard. By 2012, the Passamaquoddy Tribal Sovereign had declared a state of emergency. In 2013, we were celebrating the reopening of the Grand Falls Dam fish-ladder, which had been closed for more than two decades. Between 2016 and 2018, the two-leggeds made more mistakes and allowed more than three million gallons of black liquor to enter my veins. But I am strong, and I continue to recover. By 2019, more than 400,000 *siqonomeq* returned to my waters, and with them the *psam* and the *ahkiq* (seals) and the winged ones too. They had all started to come home.

My next celebrations were major. In 2020, the Woodland Mill replaced a 3 kilometre pipeline, and NB Power shuttered the Milltown Dam on my lower reach. The two-leggeds celebrated for me, and for the re-established $15 million fishery and its associated activities. The lobstermen had lower costs and ecologically friendly local bait. They no longer worried about introducing parasites or pathogens from imported bait fish.

As time went on, the communities of Calais and St. Stephen, buoyed by my health, developed new business and community models that were open source, technologically literate, socially inclusive, and financially productive. Profit and market success, however, are no longer central as the two-leggeds in the region now use the quantity and quality of collaboration as the core metric of success. When two-leggeds seek solutions to problems that remain unsolved they now start to build from the edges of previously unexplored and undervalued perspectives. The collective intelligence and insight of those once undervalued communities are now understood and valued.

The two-leggeds on my shores have established a new municipalism in which I thrive. This new approach puts to the forefront many truths: the truth of our global and local interdependence; the truth that social reconciliation can, in part, be created through economic justice; the truth of the power of a collective and municipal political and governmental agenda where communities own and run the services for the public good; the truth of the need to adopt cooperative forms of ownership and governance in the economic structures we create; the truth that through the empowerment of youth into leadership positions, two-leggeds can create new economic structures with better outcomes throughout the Maritimes.

The homecoming of the *siqonomeq*, the shuttering of the Milltown Dam, the return of the water and air clans, and the growing awareness and the recognition of the importance of the Peskotomuhkati Nation, the Schoodic Riverkeepers and the Maritime Social Innovation Lab (MSIL) have created a new business and community model for the Maritimes. This model is now building toward the *Skutik Kikehtahsuwakon Initiative* in which the Peskotomuhkati Nation and its settler allies place the region's woodlands and waters, both public and private, under sustainable management that is cooperatively governed.

Part One: It Begins with Land and Water

Ocean Assist Dive Log
Renelle LeBlanc

Dive Number 349
Dive location: Bay of Chaleur, zone 5 ouest, underwater kelp nursery.

It's a beautiful day in August. Early this morning, a team of six divers, including me, completed a routine dive to survey one of the newest kelp gardens located approximately eight kilometres off the Nigadoo shore in the Bay of Chaleur. The kelp in this zone surrounds one of our scallop aquaculture programs.

Since we started planting kelp nurseries, our expert divers, local volunteers, students, and marine biologists, have all seen a remarkable change in the ocean floor of the Bay, and zone 5 ouest is no exception. Our dive this morning was a success. We were delighted to report the kelp patches are in a stable condition, growing at a steady pace. We are seeing signs of recovery of native species in that area.

The team's public presentations and guided school visits are aimed at raising awareness about our activities in the Bay. They have proven to be a key element for recruitment and for fostering growing support locally, provincially, and nationally. New Brunswick and Quebec share the Bay of Chaleur and are both members of the Most Beautiful Bays of the World Network. Other members of the Network from around the world are starting to show interest in our work.

Our project was initiated when the government of Canada joined efforts with the Global Ocean Alliance to protect 30% of the world's oceans by 2030. Our small group, Ocean Assist, set out to take on that challenge and work to accomplish that same target for the Bay of Chaleur.

Since then, we have worked to form an important network participating in a Deep-Sea Environmental Aquaculture Program. Collaborators include Indigenous communities, universities, government, environmental groups, and many fishers' associations and businesses on both sides of the Bay of Chaleur. It has been an exciting time.

Each year, certified commercial divers train diving teams assigned to tasks ranging from planting kelp gardens to installing and maintaining aquaculture structures on which various molluscs are able to reproduce. Marine biologists monitor the progress of the seabed recovery as part of a university studies program.

The destructive fishing technique of seabed dragging—a common practice for many decades and an economic driver for the region—is now coming to an end because sustainable practices are proving more efficient and are helping to improve marine habitat restoration.

In the last few years nature-based development is much more on the minds of local people. This is really encouraging. We are starting to see great returns on our investments and in creating jobs.

Part One: It Begins with Land and Water

Twenty Shades of Green
Tom Beckley

The forest dazzles a spectacle of bright greens as I watch from my deck looking west from Keswick Ridge. On this spring morning, I remember this same view from over thirty years ago. Back then, the bright greens of the deciduous hardwood trees were just a few scattered flashes overpowered by swaths of dark green on the landscape. The dark greens—with their straight lines and hard boundaries—were conifer softwood plantations, the regimented legacy of a forest system that favoured fir, spruce, and jack pine to supply the province's softwood mills.

That legacy of sixty years of industrial forestry, now decades in the past, has been replaced by twenty different species in a complex, regenerating, mixed-wood Acadian forest. Back then I could scarcely imagine such a view. Let me tell you how it happened.

Early in the 2020s, the provincial government tried to force through a budget that would have used taxpayers' money to subsidize a revised Energy East Pipeline to the tune of $200 million as part of the attempted bailout of the oil industry. The scheme backfired and a confidence motion brought down the minority government.

In the elections that followed, New Brunswick was taken by *La Vague Verte*—the Green Tide. All across the north, ridings flipped from Liberal to Green. Francophone New Brunswickers delivered a narrow Green majority government, and the Green Premier got to work transforming the energy and forestry sectors.

The St. John River flooded over its banks twice in the early 2020s, as it had in 2018 and 2019. Hurricane Serena smashed into the province, blowing down 50,000 hectares of plantation pine and spruce in the southeast of the province. A smaller storm, mid-decade, devastated 20,000 plantation hectares in Charlotte County. The surge from both storms saw hundreds of coastal homes on the Northumberland Strait flooded and dozens of homes washed out to sea. Luckily, no lives were lost.

Together, these events made the denial of climate change untenable. Citizens realized that a key climate solution in New Brunswick was to build a more resilient forest. Soon, all parties in the legislature supported carbon pricing. The new revenues from carbon pricing and tidal energy sales to New Englanders allowed for a major transition in the province's forests.

The goal was to create a climate resilient Acadian Forest. At first, there was a boom in softwood lumber as plantations were liquidated to supply the hurricane ravaged eastern seaboard of the United States. But when the United States changed its building codes to prevent wood frame housing in hurricane prone areas, the writing was on the wall for industrial, conifer dominant forestry. The market had changed, the climate had changed, and people understood the forest needed to change.

A third Green majority government was elected on the strength of its environmental policies. The government passed legislation that phased out fossil fuel heating of all public buildings. The remaining softwood monocultures that once went into dimensional lumber began to heat schools, hospitals, and government facilities.

With temperature increases already underway and already irreversible due to anthropogenic climate change, the more southern, temperate, hardwood species were better suited to New Brunswick's changing climate. With softwood lumber exports in shambles, government, industry, and woodlot owners all got behind the project of restoring a highly resilient and valuable Acadian Forest.

Government silviculture subsidies shifted from supporting softwood trees to supporting mixed-wood forests. An army of student tree planters planted long-lived hardwood species—yellow birch, sugar maple, and newly discovered disease-resistant strains of beech and butternut. The new forests also included white pine, hemlock, red spruce, and eastern white cedar; all are now growing to maturity.

Creating a climate-resilient forest is a long-term project. Even now, the trees are young. The plan is to let these trees live well beyond 150 years. Short-term profits have given way to the embrace of a new forest culture. With the end of the 2020s came the abandonment of the industrial license system on New Brunswick's Crown land, which gave primary management authority to the private sector. In its place, a system of regional forest councils was established with equal representation from local communities, First Nations, the forest industry, and the provincial government.

Over 5,000 private forest owners quickly signed agreements to steward their land in restorative ways. By the 2030s, the number had increased to 20,000 owners. By 2050, over 30,000 forest landowners have now agreed to restore the Acadian forest by selectively harvesting on patches no larger than three hectares and to widely share the benefits of their private forests.

The 30,000 forest landowners enrolled in restorative forest practices agreed to the "right to roam" principle. This means the public — who support restorative forest activities through their tax dollars and carbon levies — are able to access the land of these private owners for walks and hikes, camping and foraging, bird watching and other forms of non-destructive and non-motorized recreation.

The harvest rates today on these private woodlots and on Crown land are less than half of what it had been in 2030. The province is growing forests, not fibre, and growing forests takes time. For 20 years now, New Brunswick's forests have gained a respite and are flourishing despite higher temperatures and rapid climate change. In the meantime, industrial hemp, which has revitalized the agriculture sector, supplies most of the fibre to heat public buildings.

New Brunswick is now an international leader in forest restoration, and New Brunswick's forests — public and private — have become a common heritage for New Brunswickers. The dappled landscape of twenty shades of green that I watch from my deck this morning exemplifies the changes to the forest and the changes to the ways we relate to, govern, and live with those forests.

I won't see the Acadian Forest fully restored in all its glory in my lifetime, but when I look out on the view, and when I cruise the land on my electric ATV, I can see that we are growing hope, one tree at a time.

Letters from the Future

Flood Memories and Graveyard Parks
Mary Louise McCarthy-Brandt

I come to Woodstock to visit my parents' graves at the Rural Cemetery on Houlton Road. As I drive along the Houlton Road, I see a large area of empty land. The Karnes Bakery plant is gone. The beautiful duck pond is gone. My heart quickens with a sharp breath. I wonder what has become of the cemetery on the Houlton Road.

I grew up, as they say, in Grafton, across the river from Woodstock. I have both wonderful, joyous, positive memories and searing, racially charged, negative memories of my time growing up near Woodstock. I should explain; I am a proud, sixth generation woman of African descent. The genealogy of my ancestors on this continent dates to the 1700s.

I have childhood memories of crossing the St. John River to Woodstock on the old Grafton Bridge. The bridge was built in the late 1800s. It was a fierce and intimidating presence in my young life. Thinking back to the 1960s, I would walk across the bridge as a nine-year-old and do errands for my mother. I would visit the Farmers Grocery store, and for a treat I would buy a 10-cent cone of ice cream from General Dairies.

I was so afraid of the bridge, but I could not let my fears be seen by my mother. At nine years old she trusted me to walk the bridge. I wanted to be seen as strong and trustworthy. I pushed myself to be strong. I was a highly imaginative child. When I walked on the bridge, I could hear the traffic coming long before I could see it. The floorboards of the bridge were wooden, and they rattled and shook with the passage of cars and trucks.

Now, the old Grafton Bridge is gone and so is Island Park. The Grafton where I grew up is much diminished. I miss the Park and I often wanted to visit the bridge as an adult. But that was not to be. When I was a child, Island Park was lush and green with many flowering perennials and shrubs.

Part One: It Begins with Land and Water

But the Park was flooded due to the building of the Mactaquac Dam in 1968. Progress, they called it. Now, when the water is low in the St. John River, around mid-summer, some of the remnants of Island Park, like the edges of the swimming pool, emerge from the river.

I grew up on the banks of the St. John River. When I needed to go to Woodstock, I would walk the old bridge. I walked it so many times in the 1950s and 1960s. When the dam raised the St. John River level, our families' homes were relocated and then subsequently destroyed. Our homes, the Grafton Bridge, and the Island Park, were destroyed due to floods.

Now, in Woodstock I see many changes. With the old bridge gone and with the Park gone, to enter Woodstock from the Grafton side one crosses from the top of the town at the new Grafton Bridge, a large concrete structure that replaced the old one with its wooden planking. The other way to enter the town is from the Trans-Canada Highway.

As I arrive at the cemetery, I turn slightly right, hit the gravel road, and drive towards my parents' gravestone. I stop and look at how tall the trees have grown. I am surprised they now have benches in the graveyard. There is a bench not too far from my parents' burial plot where I can sit and relax. I let out a deep breath and reflect on the time when I lived in this small town.

What happened to the graveyards? What happened to *our* cemeteries? They have become parks, funded municipally, provincially, and federally. I visit my family, and there are now benches for resting. I am excited to see the plantings of perennials and the shrubs. The graveyards have become inviting parks. People come to spend quiet time with their loved ones.

I have not been back to Woodstock since 2020. It has been many long years, and my my, what changes I am seeing. I must visit other cemeteries. I only come back now to Woodstock to visit my family's graves. My immediate family members are buried in Woodstock Rural Cemetery on Houlton Road. Why is this road called Houlton Road? Of course, if you keep driving it leads to Houlton, Maine.

I make one more stop on this day of visiting family cemeteries. As I drive back to Fredericton, I stop in the graveyard connected to St. Peter Anglican Church, on Woodstock Road just outside Fredericton. I have three pairs of grandparents in this cemetery. I feel blessed to have three sets of grandparents in one place.

Cemeteries offer a place to come and reflect. The cemeteries in Woodstock and Fredericton now offer spots for me to visit my family and to forget about the vast destruction of the flood from Mactaquac dam.

The dam brought power for many but destroyed the homes and livelihoods and communities of many others. My life and my memories of living in Woodstock will always be seared with the tragic memories of the building of the dam and the loss of a tight-knit community in the St. John River valley.

Of course, everything looks greener to me than it did in 2020. I appreciate the importance of community, and being able to spend time in the cemeteries of New Brunswick, enjoying time with friends and family that have passed on.

Part Two
Shaking the Institutions

Part Two: Shaking the Institutions

What was their Real Name Again?
Alain Deneault

No one remembers what led to the nickname. Was it out of mockery or a moment of anger or just the innocent pleasure of wordplay — one of the few amusements that remain among the ruin and disarray? Today, all over the Maritimes, the former gas stations and convenience stores where people now meet on Saturdays are commonly referred to by the anagram "Virgin." It was at "the Virgins" in every village and town where we relearned the local economy.

It was after the spectacular collapse of the oil regime that citizens with vegetable gardens and agricultural land began to gather spontaneously in these spaces so centrally located in all our communities. At the beginning, people came not so much to buy and sell, but to exchange advice and information and seeds for improving their own farms and gardens.

Against a backdrop of tensions and competing identities, which the pessimistic among us thought could descend into civil war, the starving populations found themselves, week after week, in front of the long-abandoned "Virgins" hoping to find supplies of food — potatoes, carrots, onions, and cabbage from storage, fresh vegetables in season, or grains like oats, spelt wheat, and buckwheat to store for future use.

Of the stations themselves, there is nothing left but the buildings. The signs have been vandalized by stones, which might have been thrown in moments of anger or out of sheer boredom. The bathrooms are also used for storage. The counters are where a clerk can administer what we call *duties*. Most of the goods lay on the building floor or around the building on mats and other place markers the farmers and gardeners bring in.

Rumour has it that only five gas stations still operate in the whole Maritime region, selling the overpriced fuel still being paid for by a wealthy social class that remains clinging to their steering wheels even while hiding in their walled cities. This is surprising, because much of the road system has gradually become impassable from the vindictiveness of a population hostile to urban elites. The army and long-distance truckers who work for this class remain integrated into a default Capitalism, which monopolizes the oil still being extracted off the coasts of Nigeria and Brazil and from some muddy tar sands in Western Canada. It's possible more gas stations are in operation. One man, a former shopkeeper who is not afraid to cover long distances by bicycle, claims he has seen eight gas stations in operation.

At the Virgin, accounts are kept in terms of *duties*. We refer to *duties* as the actual currency. Farmers found it necessary to use a market system to recruit the labour they needed to operate and develop their farms and garden plots. Even in cities, we grow rooftop gardens. Abandoned houses are used as winter storage facilities for the harvest. Each measured allotment of food requires a *duty*. What was initially a kind of barter has become a complex way of trading beyond just bilateral exchange. The holder of a *duty*, who, for example, receives a basket of food for the week, is obligated to provide goods or services they can offer to another party in the network. Every Saturday the clerk, who keeps accounts, ensures that the food received matches the *duties* the receiver has paid to other participants.

Although the name "Virgin" has become commonplace in our vocabulary, everyone knows it's a nickname. Occasionally, someone asks, "What was their real name again?"

For those with a psychoanalytic mindset, there were several ways to deconstruct the latent meanings of the nickname "Virgin." The years following the collapse of the economic system were so hard each family could count many deaths. We saw in the name something like a salvation, a resurrection, a new beginning. But there was also something ironic in the name. No one was *virgin*. No one was naïve considering the winters we had to go through waiting desperately for trucks carrying food to arrive at dysfunctional supermarkets. The system broke down completely when violent acts of highway robbery convinced the truckers themselves to put a halt to their dangerous deliveries.

The weekly meetings at the Virgin are now a sign of hope. They are a new form of social organization for a more austere and demanding era, but one that also heralds a new kind solidarity. We now enjoy a level of mutual aid, friendship, and meaning with a degree of intensity that none of us had even imagined possible in the old days.

Part Two: Shaking the Institutions

Truth-fullness
Ajay Parasram

Gautama Buddha once said; "Three things cannot long be hidden: the sun, the moon, and the truth." But if you were a fan of truth in the 2020s, you could be forgiven for thinking the old sage got the last bit wrong.

It wasn't just the overt lies of the American president Donald Trump, or the silencing of scientists by federal and provincial Conservative governments in Canada years earlier, or even the technocratic deceitfulness of Canada's bhangra-loving Prime Minister Justin Trudeau over things like electoral reform and reconciliation with Indigenous peoples. No, lies of that sort are the vocation of politicians and we're not rid of them yet.

What scared us in the 2020s was that society seemed unbothered by the armed neo-Nazis getting police escorts at Pride parades, the white nationalist scholars supervising doctoral candidates at public universities, and the white-chauvinist goon-gangs like the Proud Boys and the Soldiers of Odin who felt morally compelled to occupy public space with symbols of white supremacy — the Red Ensign and the Rhodesian flag. What's worse is that it was all happening amidst an accelerating climate emergency that Canadians, with one of the highest per capita carbon footprints on the planet, seemed reluctant to act on. Although things seemed bleak for those of us interested in evidence, history, and the truth, old Buddha may have known what he was talking about after all.

When the Liberal party imploded electorally, the genocide-denying leader of the federal Conservative Party forged an alliance with the anti-immigrant far-right party, which outnumbered the coalition between the Greens and the NDP that first tried to form government. Together, the leaders of the right-wing parties set about building pipelines, criminalizing Indigenous land and water protectors, hardening the border, and trampling all over the rights of other-than White people across the land.

Those years after Donald Trump claimed a broken electoral system, white nationalists north and south of the forty-ninth parallel felt secure in the continent's leadership. They left YouTube to take up an increasingly public stance with their historical fictions.

The first stand-off came in Mi'kma'ki, at the site of the Treaty Truckhouse in Nova Scotia built by grassroots Mi'kmaq people to assert sovereignty and promote meaningful discussion across cultures on questions of development in the Maritime region. Following through on the promise to build a national energy corridor inspired by the violent colonial railway, the Conservative government crushed environmental and Indigenous resistance by pre-emptively mobilizing the Canadian Armed Forces to destroy the Treaty Truckhouse, which they saw as an obstacle to their carbon-centric development.

The white nationalists were jubilant, waving Canadian flags, celebrating in the streets about how they had taken back 'their' country, and singing the national anthem like packs of drunken hyenas. The brazenness of it all was the most terrifying. Mi'kmaq people doing nothing more than practicing their treaty rights to fish were cornered by white mobs of fishers and accused of "destroying livelihoods." White mobs had started by destroying Indigenous fishing gear and their lobster catches as far back as 2020.

It was too much, even for most of the people who had delivered Conservatives their mandate. Indigenous peoples, African-Nova Scotians, settlers, and immigrants of all sorts descended upon the site of the Treaty Truckhouse and held a massive rally that was live-streamed across the Maritimes and all of Turtle Island. Hearing the words of the grassroots grandmothers and with the knowledge that Mother Earth was just a few short years from catastrophic climate disaster, the workers who had been contracted to build the pipeline highway had enough and walked off the job.

After generations of being an internally displaced workforce, the pipeline workers in New Brunswick and Nova Scotia led walkouts that left their bosses dumbfounded across the country. Big business and big unions were unprepared, and the corporate media, which had never covered grassroots politics and movements, had no idea who to blame, aside from their tired script of "foreign interests," "billionaire socialists," and "post-modern neo-Marxists."

From the Atlantic to the Pacific, ordinary people began setting up safe conversation rooms in libraries, campuses, pubs, churches, masjids, temples, gurdwaras, and synagogues. The invitation was simple: come, talk the truth, ask questions that need answers, and understand why structural white supremacy, capitalism, and colonialism hurt us all.

What unfolded was a beautiful and organic revolution that arose out of a five-hundred-year history of anti-colonial politics and a deep desire of people who had been rendered powerless by broken political and economic systems to seek refuge in the one thing that never betrayed them — the truth. Not the kind of truth that led to "alternative facts" or defensive deflections or racially fragile denial, but the kind of compassionate truth-seeking that could only come about from a genuine desire to gain knowledge through introspection and attention to diverse histories. To talk truth was to commit to a process of shared learning among equals.

The bosses offered to double the wages. Nothing. They tripled them. Still nothing. They grew desperate in their attempt to convince working people to put their shovels back in the ground. Some people went, but not enough to get the pipelines back on track.

Settlers who had historically felt fragile in activist spaces began to turn up in earnest to those safe conversation rooms. People asked questions, shared food, and made connections. No one shouted anyone down for being ignorant. People came to learn, not to fight. There were conversations about colonialism and the gender spectrum and why the idea of "self-interest" is a culturally relative value rather than one that is universally true across all cultures. Through these conversations, we learned that we already had the means to address catastrophic climate change and organized to force the hand of governments through direct actions and mass movements.

It was these spaces for conversation, sharing, and learning that sowed the seeds that would ultimately crack the foundation of the neoliberal university system and transform the notion of "higher learning." Meanwhile, the white nationalists were outraged. They turned their outrage on artists, activists, scholars and all the rest, whom they accused of lying about the past to trick white people into hating themselves.

But hate had no place in those conversation rooms, which were guided by an acceptance that the past and the future were out of our hands, and it was only in the present that actions count. Like a raging fire deprived of oxygen, the white nationalists found that people were tired of being lied to and were motivated to save their communities, their societies, and the whole world.

We never fully resolved racism, of course. But the rejuvenation of truth and compassion helped people to understand the solidarity they not only had with one another but also with the Earth itself. This was an essential lesson, which helped us prepare mentally and materially for the waves of climate refugees arriving in the last decade.

It was a place-based solidarity and grounded ethics, which was not new. Societies all around the world have long possessed deep knowledge of place before the colonial encounters sought to shovel that knowledge into the steam engine of endless accumulation and development.

In a world struggling to overcome the deep scars of greed, we are all still learning to live truth-fully with proper respect for the land that sustains us. Despite the challenges we face, there is something beautiful in sharing a world that recognizes the value of truth-fulness and views compassion, not competition, as the essence of human nature.

Part Two: Shaking the Institutions

Who Saved the World? Girls!
Lauren R. Korn

I hope this letter finds you well. Me? I'm settling back into life in Fredericton. I've been away, out west, and like most places I've called home, Fredericton seems to have both changed and stayed the same. Neon flowers that once flourished in front of Maritime bungalows are flourishing still, and the coastal humidity I once cursed still hangs in the air, tethering my body to itself.

Sticky skin aside, I'm writing to tell you that I recently accepted a position with New Brunswick's Centre for Women's Health and the Environment, a not-for-profit organization established in the early 2020s that operates as an interdisciplinary learning and activist space for residents of all genders and identities, a space to consider the ways that physical and mental well-being are approached and understood in relationship to natural spaces both provincial and global.

Shortly before the Centre was established by a group of community partners, all of them women, discourse surrounding the climate crisis had gained momentum. It was becoming apparent that to halt the rapid deterioration of the planet's ecosystems, including our human place within them, women's experiences of labour (read: the workforce), domesticity, family planning, contraception, the sustainability of "feminine care" and hygiene products, and accompanying mental health concerns needed to be at the forefront of the conversation.

Members of the Centre for Women's Health and the Environment include researchers, politicians, mental and physical health professionals, economists, activists, filmmakers, poets and writers, permaculturists and farmers, small business owners, and others. They all work to help centre, and thus equalize the narratives of women and femme-identifying persons and to provide area residents with information about provincial services that can help them untangle weighty life decisions; that is, how they can best claim agency in their own lives without contributing to population growth, to the world's climate crisis, and to its associated social inequalities.

For instance, the Centre works with federal and provincial officials to champion a Guaranteed Annual Income (GAI) that ensures previously unpaid domestic labour—like housework and child-rearing—is recognized for its contribution to the economy and compensated accordingly. Domestic labour inherently upholds capitalism's power structures and must be given financial support to maintain healthy families and communities. Members who took part in the GAI initiative are currently working to increase female representation in business and banking and are particularly interested in addressing the gender wage gap and pay inequity, discrepancies that still pervade our provincial economy.

The Centre has also partnered with schools in both urban and rural areas of the province to implement a feminist and environment focused sexual health education system. The Centre works in tandem with provincial hospitals and clinics to provide timely and safe abortions, as well as contraception and access to mental health care both prior to and following every patient's consultation or procedure.

If you're thinking this is an extreme contrast to the New Brunswick of 2019, you're absolutely right. In 2019 we saw attacks on reproductive rights across the province with funding restrictions on abortion services. Schedule 2 in provincial Regulation 84-20 of the *Medical Services Payment Act* limited abortions to hospital settings and was in violation of the *Canada Health Act*. This regulation made reproductive and sexual health services increasingly difficult to access for women and transgender patients in lower income brackets and outside city centres.

The Health Minister, who remained ignorant of reproductive health issues and intransigent on the provincial restrictions, resigned his position shortly before the Centre took up residence in Fredericton. This resignation and the work of Centre have done this river community and the province a lot of good.

When sexual health and reproductive rights were given the space and consideration they deserve, the physical and psychological wellbeing of New Brunswick's entire population improved. Now, clinics around the province are thriving—as are their patients. I feel a particular, personal stake in this conversation. As a woman, yes, but as a woman of a certain age, thinking now about starting a family—about *whether* to start a family—and what that decision means for population growth and a warming planet.

With both conviction and empathy, I feel the freedom to choose, just as I felt the freedom to choose prior to my first tenure in Fredericton.

The mere fact of my body means that when I first moved to this sleepy New Brunswick town in the summer of 2017, learning about how my body would be received and treated in public spaces like doctors' waiting rooms and offices became deeply important to me. As someone who grasps tightly her relationship to the natural world, the decisions I make in relation to my body's place and to my body's production *in* it are equally important.

I am moving forward in my role at the Centre for Women's Health and the Environment with these freedoms and responsibilities in mind. I move forward with the certainty that New Brunswick will continue to make educated and community-supported decisions for a healthy and sustainable future.

Electoral Reform
Victoria Clowater

Even though New Brunswickers had been calling for electoral reform for years, we didn't fully realize how much it would benefit our province until after we achieved this goal. After switching from our first-past-the-post system to proportional representation, we created a New Brunswick for everyone living here.

New Brunswick's old electoral system was meant to represent residents in all areas of the province. Yet it seemed that, once elected, MLAs put the interests of the party and the powers-that-be before the needs and interests of voters. Rather than electing representatives, voters elected political parties that did what was best for the party itself.

Voting choices were limited. Although new parties arose from time to time, only the red and blue parties ever held power. True representation and democratically shaped progress was out of reach as long as our province flipped back and forth between the same two options. Voters anticipated change but always elected one of the same two parties to govern us. It was often the case that the leaders of the two governing parties previously had long careers with New Brunswick's most powerful companies or associated themselves with the province's financial elite, ensuring that business interests remained the primary focus of government concerns.

After the 2018 and 2020 elections the problems with the voting system became obvious. In 2018 the Liberals received the most votes but not the majority of seats. The Green Party and the People's Alliance won only 6 of the 49 seats, 12.24%, but collectively they had received 24.26% of the popular vote. The NDP party received 5% of the popular vote yet won no seats. The support these parties had in the province was not equitably reflected in the makeup of the provincial legislature.

After this pattern was repeated in the 2020 election, New Brunswickers decided this was enough and launched a grassroots campaign to advance electoral reform. In 2030, after a decade of growing support from people across the province, the newly elected premier followed through on a promise to implement electoral reform.

The 2034 election was New Brunswick's first using proportional representation. One of the most notable election outcomes was the 84% voter turnout—the highest voter participation in decades. Voters finally believed their votes would count, no matter whom they voted for. The Greens and the NDP, always on the margins, were leading the polls. New Brunswickers were no longer reluctant to vote for their preferred parties because their vote was no longer "wasted."

With the election of these parties came new, innovative, and social justice-minded policies. Regulation 84-20, Schedule 2 (a.1) of the *Medical Services Payment Act*, which illegally restricted access to abortion in the province, was repealed. This led to greater access to abortion services across New Brunswick. It was a major milestone for reproductive rights in the province, which had been ignored for decades by successive Liberal and Progressive Conservative governments.

The newly elected government introduced greater support for people with disabilities. New legislation provided the support they needed to live in their preferred communities, adequate home care, and full coverage for their needed medical supplies. Trans New Brunswickers were provided with greater access to their needed healthcare wherever they lived rather than being available only to those living in the three major cities. The province worked with the New Brunswick Medical Society and Dalhousie University's Medical School to implement better training for doctors in transgender medical care, which helped break down systemic healthcare barriers for trans people.

The new government invested in long overlooked mental health and addiction services. Funding was finally allocated to increase access to these services and make them accessible to everyone in the province. Instead of penalizing drug users, the province adopted a harm reduction strategy and provided support to anyone asking for it. The number of detox beds was greatly increased, and the wait between detox and rehab was eliminated for those seeking recovery.

Education reform was another priority of the new government. French immersion was introduced in schools across the province, addressing a gap that had long plagued rural anglophone schools. The updated curriculum included previously overlooked aspects of Canadian studies such as Indigenous history and the history of slavery in Canada. The new curriculum became a model for other provinces. The revolutionized sex education curriculum focused on harm reduction, empowerment, and equity for gender and sexual diversity.

Inspired by the changes that took place during the four years when New Brunswick was led by its first proportionally representative government, new political parties were formed in time for the 2038 election. More New Brunswickers than ever found representation among these parties, and many people who would never have dreamed of running for office decided to go for it.

The most diverse array of candidates ever was elected to the provincial legislature in 2038, representing seven different parties. Despite concern that the new electoral system would lead to unstable governments and frequent elections, in practice it led to government representatives committed to working cooperatively, making positive changes for New Brunswickers rather than seeking power for their political party.

Cooperation was not limited to the elected representatives. The new government worked collaboratively with First Nations to repair and transform the relationship between the government and the Indigenous peoples living here from time immemorial. The government finally acknowledged its obligations under the Peace and Friendship Treaties of the 1700s. Indigenous New Brunswickers were recognized as the official stewards of the land, and no more natural resource projects were implemented in the province without their express and un-coerced consent.

When New Brunswickers finally came together, demanded reform, and voted for progress rather than for the most strategic candidate, the province was able to break free from the old cycle of volleying back and forth between blue and red. People no longer felt their province was an oligarchy ruled by a small group of powerful people with personal connections to party leaders. New policies made New Brunswick a wonderful place to live, not only for those with money and power but also for everyone who called it home. By switching to a proportional representation electoral system, we created a New Brunswick that uplifted all people. We transformed our systems, our relationships, and our lives for the better.

Part Two: Shaking the Institutions

The New Brunswick Youth Parliament
Arianne Melara Orellana

Today is a special day; it's my daughter's fifth birthday. I find myself wrapped around with feelings of joy and comfort thinking about the bright future awaiting her right here in New Brunswick.

Eight years ago, my partner and I chose to stay in New Brunswick to start a family. The events transpiring since have shaped the province in a way that validates our decision to continue living and working here. Our daughter will grow up in a province that prioritizes the voice of youth. After the province introduced the first New Brunswick Youth Parliament (NBYP) things began to look up for youth living here.

The NBYP has 49 young persons representing all 49 ridings across the province. They range in age from 17 to 24 and come from diverse backgrounds representing Indigenous peoples, Francophones, Anglophones, and Newcomers. New Brunswick's *Official Languages Act* has been expanded to recognize the languages spoken here before colonization. The formal meetings, written communication, and even social circles and side conversations of NBYP use New Brunswick's five official languages: Mi'kmaq, Wolastoqewi, Peskotomuhkati, French, and English.

The change was possible because the province also introduced a high school immersion program for students to learn all the official languages. Now, graduating students form part of the new generation of multilingual citizens with an in-depth understanding of the history linked to these languages, the cultures and identities they represent, and the importance of their conservation and use to benefit all New Brunswickers.

The NBYP was created by the Legislative Assembly and the New Brunswick Child and Youth Advocate agency in partnership with many provincial organizations. It is now part of the new Ministry for Children and Youth Affairs. The Youth Parliament tackles local issues affecting young people and has increased the participation of young people in civic affairs from all walks of life. The issues are brought to the table and motioned by young people themselves at a Bi-Annual General Meeting.

The inauguration of the Youth Parliament helped affirm the value and potential of young people in the province. Since its creation, youth have been more involved in formal politics, more informed about local priorities, and more knowledgeable and aware of both their power and potential as citizens of New Brunswick.

The first Youth Parliament initiative passed by the Legislative Assembly and become law was an amendment to the *Election Act* lowering the voting age to 16. This has transformed voter turnout to an all-time high of 87% among the younger generation. Youth are inspired to make a difference because they see their role in making decisions for the future, and they know their voice is validated and heard. As early as age 14 youth are informed about the vital role the Youth Parliament plays in the province as well as the importance of fulfilling their duty as voters. This knowledge has increased positive mental health and wellbeing among young people. They develop a sense of belonging filled with purpose and know they can share their talents and perspectives in a collective, democratic way.

New Brunswick has begun reaping the benefits of the Youth Parliament. If you take a snapshot of the Legislative Assembly today, you see a picture of the province's diversity. We have elected younger, culturally and linguistically diverse representatives from a range of gender and sexual identities. No longer are old white men dominating New Brunswick's legislature.

It's encouraging and refreshing to know that after so many years of gender and racial exclusion, my daughter is growing up in a society that values diversity in all its shapes, ways, and forms, and where politicians of all stripes value the voice of youth. This transformation has created an intergenerational shift in the minds of New Brunswickers that promotes innovation and change. The presence of young people of diverse backgrounds in the legislature has engaged youth in politics, decreased their apathy, and increased their civic participation as concerned citizens.

As New Brunswick realized it needs the talents, skills, and perspectives of young people to grow, excel, and flourish, the government opened worlds of opportunities to international students in the province. Any international student completing a four-year degree was nominated to Immigration

Canada for permanent residency upon graduation. The program allowed participating students to work here after graduating. At first, people thought the graduates would go on to other provinces after receiving their permanent residency, but this was not the case. They stayed, started careers, and formed families after finishing their studies.

The creation of a collaborative program that matched students with paid internships in the public and private sectors helped graduates envision future careers in New Brunswick. All students, including international students, now graduate from universities and colleges with careers already in mind, making them ready to enter the provincial workforce and help our economies grow and thrive.

The internship program formed part of a professional, provincial hub designed to equip young adults with training in facilitation, communication, and other leadership skills. Previously, the organizing and facilitation of conferences and other events in the province were often contracted to professionals from outside New Brunswick. Now, more locally trained leaders are taking the stage to provide organizational and facilitating services, acting as role models to others early in their careers, and sharing their expertise shaped by growing up and living in New Brunswick.

The impact of making youth a priority has resonated in communities around the province. Community based organizations and boards of directors now often include individuals under 30 years old who are responsible for leading on social change initiatives and at the forefront of key governance decisions. At one time, we thought our daughter would have to move to a bigger city in another province to have leadership and empowerment opportunities. Not anymore. With its embrace of diversity and investment in youth, New Brunswick has a fast growing population (almost a million!) and greater opportunities for youth right here are rapidly increasing.

Letters from the Future

The Revolution of a New Generation
Alicia F. Noreiga

"Black lives matter! We want peace!"

The chorus of hundreds of protestors is etched in my memory. I close my eyes, remember that remarkable day, and I cannot help but be fascinated with the transitions that occurred over the years since that time.

I marched among the large crowd of demonstrators, wearing facemasks and keeping an acceptable distance from each other, amidst the COVID-19 pandemic. Carrying my placard, I stood in solidarity among the hundreds of people of different races, determined to disrupt the intersecting levels of oppression that have covertly and overtly disadvantaged people of colour. The huge number of White supporters present mesmerized me, acknowledging by their presence the existence of systemic supremacy bestowed upon their race over others.

Most of all, I recall a feeling of hope and the thought that came to me as I stood before the young crowd. "Was this the beginning of a new generation who would change the course of events in a province that has in the past routinely segregated Black people and erased an awareness of their lives?"

Now, it is clear, I have witnessed a metamorphosis of a society that was once ignorant of the many social ills that systematically isolated and prejudiced persons belonging to minority groups. A community has now emerged with a nucleus that drives systems to precipitate social justice and equity. It has been nothing short of an absolute privilege to witness this revolution in New Brunswick. People now willingly acknowledge their shortcomings and take measures to ensure the establishment of an inclusive community where equity and social justice are more than rhetorical mantras—where they are embedded deep in all the province's systems.

We have come a long way — a long way from the time when Black histories, Black contributions, and Black discussions were entirely avoided in the classrooms and education curriculum; a time when Black students grappled with feelings of detachment, conscious not only of their difference but also the lack of recognition of their Blackness brought on by silence about Black contributions. New Brunswick's traditional history impeded the understanding of Black connections to the province.

I am fortunate to have witnessed the close bonds formed between Black and Indigenous communities. They have worked to support each other in unearthing buried histories that will serve as catalysts for a brighter future. This comradeship has heightened activism. Once silenced voices have gained strength as they press for positive change by initiating discussions with stakeholders and policymakers

Today, as I kiss my grandson when he leaves for school, I am comforted by the knowledge that he feels a sense of belonging, that he feels a part of his school and community. He often returns home from school elated about things he learned that day, boasting of the significant contributions made by New Brunswick's Black residents. His diverse classroom consists of students of all ethnicities, races, sexualities, and abilities, yet everyone feels equally empowered to contribute. Individual uniqueness is embraced as a nexus of both school and community.

Some of these changes emerged from the Department of Education, which facilitated curriculum modification projects that rectified the erasure of Black awareness and Black histories from the education system. The province's universities successfully incorporated Black awareness in their teacher training programs, equipping teachers with the knowledge and skills necessary to integrate Black issues in their pedagogies. All education institutions, including universities, recruited Black educators and support staff into their institutions, which provided mentors for their Black students.

We have certainly come a long way in New Brunswick from the time when immigrants and racialized groups were alienated by the absence of support for cultural diversity, and which pushed them to leave the province they could have otherwise grown to love. Over the past five years, New Brunswick's population has been steadily increasing. This, in many ways, can be credited to the province's efforts toward multicultural awareness and support. The presence of diverse food, music, and celebrations have attracted immigrants and locals who, with a growing sense of belonging, have chosen to stay and work towards making the province a vibrant, exciting place for all residents. Now, the urban centres of the larger provinces have lost their allure for New Brunswickers.

Jobs and other opportunities are now equally available to all people without regard to their race or culture. All people are respected and considered equal. The province has become known as a place with the opportunities for all to contribute and build a livelihood. Out-migration, which previously led to an increasing older population, has been reversed. A steady influx of returning New Brunswickers and new immigrants has created a better demographic balance between young and old in the province.

Ours is now a vibrant, innovative, diverse population that embraces differences and celebrates its cosmopolitan composition. Ours is a population that envisions and is devoted to a sustainable New Brunswick. Through this dedication, the province has achieved a balance of Indigenous and contemporary practices that provides effective economic development without sacrificing environmental sustainability.

It has been eighteen years since I left my home and settled in New Brunswick. I am proud to belong to a province that embraces my uniqueness, my blackness, and my foreign accent. This is my new home. Here everyone works together toward preserving their community, eradicating racial prejudices, and establishing equity and social justice for all, regardless of differences.

We have not reached a utopia; there is still a long way to go to ensure equity among all races, genders, sexual orientations, and social classes. However, the progress we have made over the past years reflects a province committed to a better future for all. We are definitely on the right path.

Part Three

Welcoming Newcomers and Coming Home

Part Three: Welcoming Newcomers and Coming Home

They Came in Waves
Jael Duarte

The 2020s were years of unexpected demographic change in New Brunswick. Young people stopped leaving for jobs out West, and those who had gone west came back to the province in waves. But that was just the beginning.

First came the New Brunswickers who had left for work or education and now came back to escape expensive cities and precarious jobs. Next came the elderly, fleeing the dissolving suburban dream in the drought-parched lands of the US Southwest and the storm wracked regions of the Southeast and Gulf Coast. Unexpected, however, were the people internally displaced from across Canada, people now dodging forest fires that engulfed their towns and floods that destroyed their homes. In addition, climate refugees came from all over the world seeking safety and solace.

As a province, we embraced the advice of the United Nations Climate Change Group. At first, "climate refugee" was not a legal concept, but then a legal framework was developed for refugee protection that included people fleeing conflict and famine brought on by climate change. We welcomed the waves of people flooding into the Maritimes, the modestly wet, temperate, most affordable, and liveable region of Canada.

Early on, it was the youth who chose to stay at home as the broader economy inexorably shifted away from oil. The shift, first slow then fast, led to economic and demographic changes. As the oil industry failed, the allure of jobs in the Alberta tar sands faded. Employment in the industry became increasingly dangerous, low status, precarious, and poorly paid. Young people in New Brunswick found work building passive solar energy homes, retrofitting old houses, installing solar panels, and building a local, renewable energy electricity grid. More and more the dream of New Brunswickers became to stay home.

Students were attracted to low tuition, small classes, and the educational, training, and employment opportunities. Families came for affordable housing and safe communities in which to raise their kids. People came for a better way of life, with cooler summers, no water shortages, and a less crowded environment. Retirees fled the cities in central Canada, the hot summers in Arizona and the flooded coastlines of Florida. Many had little choice. New Brunswickers as well, were forced to move inland to escape rising sea levels. From across the globe, climate refugees arrived in New Brunswick, Nova Scotia, Newfoundland, and, surprising to some, even Labrador. The Atlantic Provinces became a destination for internal and international migrants looking for a good place to live.

New Brunswick led in Canada, and Canada led at the United Nations in drafting, signing, and applying a new Climate Refugee Convention, creating the legal framework and administrative tools for responding to climate refugees. This allowed the province to build effective public policy for dealing with this new emergency. Civil servants came from all over the world to learn from New Brunswick's experience.

Like few other places on the planet, New Brunswick welcomed and embraced people fleeing climate change. It was rocky at first, but New Brunswickers soon saw the benefits. Paradoxically, the solution to a major demographic problem came from climate change. The newcomers were a godsend for helping the province sidestep a severe demographic imbalance of its aging population.

Low cost, safe housing and the opportunity to make new lives attracted many newcomers with new energy, ideas, languages, and cultures. We embraced this diversity. Our cities became more cosmopolitan, our towns and villages were revitalized with a new diversity. Our settlement agencies were well funded to support newcomers, help families to settle, and especially empower women to fully participate in community and economic life.

A new wave of feminism emerged from all this participation and leadership. Men and women worked together increasingly conscious of their equal contributions. Communities adopted new attitudes toward raising children and the importance of personal care work. Childcare and eldercare became valued and well paid. Teachers and preschool workers were fully valued. The immigration system was changed to allow newcomers to bring their parents and siblings so that extended families could more easily share care work.

The new arrivals began creative careers, opened new stores, began light manufacturing operations, filled service jobs, worked as farmers, in healthcare, in public service, and in the professions. Enterprise and employment prospered in all fields.

For decades, social scientists, economists, and politicians had predicted a demographic tsunami of aging residents would hit the Maritime provinces. In the end, the solution to caring for an increasing proportion of aging residents with increased health care costs and a declining provincial tax base, was not more neoliberal policies; for example, lower corporate taxes, subsidies for big business, more export-oriented resource extraction, and ever greater provincial dependence on the globalized economy and global culture. The solution was in welcoming and supporting people to move to New Brunswick from across Canada and around the world and for them to make New Brunswick their home.

Letters from the Future

Go East, Young Woman
Leland Wong-Daugherty

I'm excited for next week. On Wednesday I'll be heading to the Land Services New Brunswick office in Fredericton. My family is hoping parcels of land will still be available outside Fredericton city limits. I don't mind where the land is. I'm just hoping the land available will be near water.

I was born in New Brunswick, which means in the first six months of the year I can choose and apply for a parcel of land. During the second half of the year, parcels of land are allocated to newcomers.

Nearly all my childhood friends have come back to New Brunswick to a get a parcel of land—some from Alberta, others from Ontario or British Columbia or the United States. As New Brunswickers, we have a right to our own small parcel of land. It's a birthright. The land is free. Newcomers can apply as well.

Seven years ago, legislation allowed a pilot project for land allocation in New Brunswick. The project started on a long abandoned one hundred acre farm near Zealand. The property was divided into homestead parcels of one hectare.

My younger sister Amanda applied for a homestead when she turned eighteen. She got the first one. My parents were horrified. She had been accepted at Ryerson in Toronto. They wanted her to leave, to go off to university. But she chose to stay in New Brunswick. Seven years later, she's still here.

Now, nearly 3,500 acres of abandoned farmland has been purchased by the government, subdivided into parcels, and given away as homesteads. Many people want a homestead, and my sister is having the time of her life.

Amanda always loved outdoor life. She has rejuvenated an old apple orchard. Her raspberries and blackberries are incredible. She plants a huge vegetable garden each year. She added a pond to her homestead and keeps a flock of ducks that run at you when you visit. She keeps goats for milk and makes delicious cheese.

I wish I had started when she did. We are competitive that way. My kids want to go to Amanda's place on the weekends. Our parents like visiting as well, even though they still pester her about university. Amanda might go one day. But she met a guy out there on her homestead and, excuse the pun, they're two peas in a pod.

They say New Brunswick now has the highest number of small farms in the country. People are beginning to call it Canada's "garden province." Tourists come just to buy the food people have grown on these homesteads: exquisite berry preserves, pungent gourmet cheeses, fragrant fruits, and delicious vegetables. For the first time in decades, the population in New Brunswick is inching up. It might double in ten years, I heard. We now have 1,300 mini-farm homesteads. That's a lot of goat cheese.

Researchers say the gardens are good *carbon sinks*—something about *humus*. Not the edible *hummus*, mind you, which lots of people here make and sell, but soil *humus*. Plants pull carbon dioxide from the air. Some of this carbon is transformed and deposited into the soil with the help of fungi. It's a process of humification, which traps carbon. New Brunswick's garden homesteads are fighting climate change by naturally sequestering carbon.

Plant-Heads—who prefer to call themselves Plant Specialists or Farm Extension Agents—visit the homesteads across New Brunswick to encourage and help establish gardening techniques that promote carbon sequestration. In fact, Amanda already has full-time employment as a professional Plant-Head, much to our parents' chagrin.

Like many of my friends, I was planning on moving my family out west. But, now I'm having second thoughts. I can get land here and there is so much going on. Folks are arriving from all over the world looking for a homestead. Some have long hair and beards, but most don't. Many are newcomers from places like Colombia, Denmark, Germany, El Salvador, Jamaica, and Russia. They are all looking for their little piece of green heaven right here in New Brunswick.

The province has done many things to support this rural renaissance. It acquires and gives away abandoned farmland. It doesn't tax the homesteaders. Instead, it charges a small tourist-tax. The tax hasn't dissuaded tourism, and the press coverage of all of this has put the province on the map.

I always knew New Brunswick had a gift for the world. With the allocation of abandoned farmland into small parcels for rural homesteads, we've found it—right under our feet. So, we'll likely do what Amanda did and stay East.

Part Three: Welcoming Newcomers and Coming Home

How Far We've Come!
Sagda Elnihum

How far we have come with the treatment of immigrant women! Living conditions of immigrant women in New Brunswick, in the Maritimes, and in Canada have greatly improved. Although injustice continues to exist, New Brunswick is now a province of progressive innovation. The province in which I arrived, is long gone, and with that passing a new era of fairness has dawned.

I arrived as an immigrant from the Middle East. I know well the hardships people like me once endured. I used to dream about the day when we would all be treated equally and hoped that day would come soon. I used to hope that we could create fair employment regulations, to be evaluated for our skills and not our gender or race. It worked. It took considerable time and effort to achieve but now, finally, things are improving.

It's almost unbelievable. Every day, I hear about new changes. Almost, almost, it feels like the equality we have been waiting for has arrived. I look back and see we worked hard, with many setbacks. But we never gave up, and I am proud to know that even when times were most uncertain, we kept moving. Now, more than ever, I feel we must keep moving forward.

Thanks to all the men and women who spent so much of their lives defending immigrant women's rights, we can all live in harmony, and I hope lives will continue to improve. One thing is certain in New Brunswick—no voice is too small. Our voices were heard and so were our demands. I am proud of what we accomplished as a community. What we thought impossible—fair employment, equality, ending racism, respect for each other—is now a reality.

I am proud of the young people and students who accomplished so much. They will carry on with our dreams and never stop dreaming. Dreaming is not enough though; we must also work hard to attain more rights. It takes hard work to accomplish such dreams. As immigrant women, we must equip ourselves with knowledge, and we must achieve our goals.

Like our generation, young people now have their hopes and goals. I believe they will accomplish theirs, like we did ours. They have proven to be more aware of the struggles of immigrant women than we were when we arrived in the province so long ago.

It feels good to look back at the amazing work of our predecessors that we carried on and bore such fruit. In the past, there were many struggles, but I am proud to say, as a community in New Brunswick, we never gave up. Our province has become an accepting and kind place.

Part Three: Welcoming Newcomers and Coming Home

What's Changed? A Lot!
Sara Taher

Today marks a decade since I arrived in New Brunswick. Things have changed a lot since then. Today, I look around and feel happy with how my life has turned out here in this small green city in this small province.

Fredericton has become a multicultural city. I come across other immigrants all the time. Canadians in New Brunswick now know more than ever about other cultures and are beginning to understand and accept differences. Immigrant retention, historically a challenge for New Brunswick, has become easier. New Brunswick has emerged as a good destination for newcomers.

Things worked out for the best, which makes me happy. For example, on television I see more positive news and information about other countries and cultures. My neighbours and I chat about Arabic, Indian, and international movies and talk shows translated into English. To think, at one time, none of this existed.

What's more, permanent residents now have the right to vote, which was not the case when I first arrived in New Brunswick. Giving permanent residency holders the right to vote was a smart move, helping reshape New Brunswick for the better and making the province a more inclusive place.

Last week, I had a job interview. I got the job. I am excited about my new role and happy that I no longer must exert extra effort to prove my competence and experience. In the past, some people believed women wearing a hijab could never be good engineers. How times change!

My side business has also grown and is now one of the biggest modest wear brands in Canada *and* the United States. I operate my business from the heart of Fredericton, the same city where ten years ago, I felt so misunderstood that I thought about going back home every day. Not anymore. New economic policies have helped attract tech companies to the province. I work at the three-year-old Smart Village where the big companies have offices. In the past, if you wanted to work in tech, you had to relocate. The Smart Village also offers internships and training opportunities for new university graduates.

Fredericton has become a greener city. A biotech team at the university led the way in developing an environmentally friendly pesticide. No longer is the province sprayed with harmful chemicals. New Brunswick residents have begun to install water-recycling technology. Many homes and businesses now recycle most of the water they use. People water their lawns and wash their cars with no problems or concerns because recycling technology reduces water use to such an extent.

So much has changed: more and better jobs, a better society for immigrants, a greener approach to living. I can see now that I made the right decision by moving to New Brunswick.

Part Three: Welcoming Newcomers and Coming Home

A Letter from El Salvador
Carlos Morales

My dearest, I hope that when you receive this letter you will be in good health. I long ago lost the habit of writing letters. I said the same thing to your grandpa, when I decided to spend my retirement in El Salvador and told him about the possibility of living in the house he has outside the capital.

Of course, writing poetry is something else. But poetry is not a letter. While I always have paper and pen to write poetry, when it comes to knowing how family and friends are, it is much easier to use a cell phone or a computer. But here is a letter for you.

The banana trees at grandpa's house have grown a lot. They give a great harvest. If you could see how heavy the big bunches of bananas are! But grandpa always hopes I won't keep the fruit just for myself. He has this thing about not wasting anything. I don't know if I can take care of what grandpa planted on his land. I want to continue writing and publishing another book of poetry. It seems your uncle might spend more time in El Salvador, and he will lend your grandpa a hand.

Do you remember when I told you about my childhood? Yes, my first job was when I was ten years old picking coffee with your uncle. We got up early to go to the farm, bringing the lunch that my mom, your grandmother, had made. The farm belonged to our godfather, but we didn't get any privileges as godchildren. We still had to make do with the place where they told us to pick the coffee. With a lot of effort, we each picked an *arroba* of coffee—a sack of 25 pounds.

Your uncle always loved the land ever since he was young. But your grandmother had no patience with him. I don't remember how he got rabbits to raise in the house where we lived. Your grandmother didn't like that she had to take care of the rabbits while your uncle studied agronomy at a private university.

When we arrived in New Brunswick, we learned that people also go to the countryside to harvest the season's crops. Your uncle took the opportunity to start work by harvesting vegetables with your other uncle. Your grandma may have remembered when we were going to pick coffee in El Salvador. With time, your uncle decided to study agronomy again, but he had to go to Nova Scotia to get his degree. He surprised me, at least, when he bought a farm in New Brunswick. And to think that as children we worked on the farm of our godfather. I used to tease your uncle by telling him that now that he had a farm he had become landowner.

Your mother tells me that in the summer you will all go again to the plot of land your uncle still has to celebrate Victoria Day. He and his farm! That is where all your cousins grew up. Your uncle enjoyed raising his cows, sheep, goats and chickens and harvesting his vegetables and fruits. He managed to have a good business selling meat. Eventually he had problems and expected some help from the provincial government, but the help never came, and he had to sell the farm. But he kept the plot of land I am telling you about.

It was in the years after the pandemic in 2020, when food became so expensive, that your uncle began to grow food on his small plot. Well, not just him; no, he and many of his friends in the Latino community in Fredericton. He needed help with the harvest; it was very big. Neighbours helped, and it was all reported in the news. Since then, there has been no shortage of help for our friends and their friends.

You looked so pretty when you went to pick strawberries and tomatoes, and harvest apples. When I saw them in your basket, I remembered how I used to go and pick coffee. Your aunt still enjoys making tomato sauce, strawberry jam, and apple jelly. After the pandemic, I had to be careful not to leave you alone with your basket because you would eat the fruit and tomatoes and then your tummy would hurt. Your dad tells me that now you help your aunt make the tomato sauce, but not the jam and jelly because you hide from her and eat the strawberries and apples.

Well, how happy I am to write you this letter and to know that your uncle continues with his projects and his garden on the plot in the country that is available for his community. His neighbours also donated part of their properties for gardens so there would be more local food products.

I must go, my dear. Take care of yourself; don't eat too many pieces of the fruit your aunt is preparing, because your belly will hurt later. I hope you enjoy being together with your mommy and daddy on Victoria Day.

     Your uncle, who loves you, and thinks of you a lot.

Part Three: Welcoming Newcomers and Coming Home

We Came, We Saw, We Helped Build the Economy
Lisa-Gay Taylor

As we sat on the patio over Saturday afternoon tea, we reminisced about the province and its development. For a group of immigrant women, it was not our usual conversation over tea. We were baffled remembering the news stories from decades ago—New Brunswick's rate of population growth turning negative, the number of deaths exceeding the number of births.

Yes, there was a time when death rates were higher than birth rates. The province's population growth was not coming from within, and migration from other parts of Canada was not enough. All the provinces were facing demographic challenges, but New Brunswick's was the most acute.

People came out east to retire, looking for a place to live out their last years. In 2020, 100,000 people in New Brunswick were nearing retirement. It was a demographic crisis, and the province turned to immigration from around the world. Immigration was the only way to ensure the province would have enough workers to meet its labour demands.

Immigrants like me had only come in the hundreds annually until 2020. That year, just 31,000 immigrants lived in New Brunswick, 18,000 of whom were employed full time. At the same time, the number of people available to fill jobs across the province was in a steady decline—transportation, manufacturing, primary and service industries were all facing a crisis.

So, it was people like me and my friends, now having tea, left our homelands to seek a better life in New Brunswick. We travelled from afar to this province for many reasons: some were family-class migrants closely related to Canadian residents already here; some were economic immigrants, either those with Canadian experience, skilled workers, temporary foreign workers, businesspeople, or participants in pilot programs; some were people accepted as immigrants for humanitarian or compassionate reasons;

some came as refugees who were escaping persecution, torture, or cruel and unusual punishment.

Some of us came to join our spouses and families, some to find job opportunities, some to study, some to seek sanctuary. Some of us came because of overcrowding in our home countries. Often we did not come alone; we brought our families, money for settlement, valuable skills, education, and investments. We came with the energy and talent and money that made New Brunswick what it is today, the most economically important province in the Maritimes.

At first, some people in the province did not like the idea of people moving in from other places. Some argued that immigrants were harmful to the province and the economy. Some devalued our work and our contributions and were jealous of the upward mobility immigrant families often achieved.

We faced many challenges. We faced repeated rejections when applying for jobs. Too often, we worked for minimum wage, at levels far below our qualifications. Sometimes, we were forced to work longer and longer hours. We had to navigate the Canadian entrepreneurial ecosystem. We sought funding and programs. We adapted to the weather, to the language, to culture shock, to social isolation, and to the credentialing systems. Even though our education and qualifications were more likely better than most others in the province, as immigrants we were rarely hired for white-collar jobs.

But we worked hard to build the economy and to achieve long-term goals. We made this province our home, and we saw it as our role to make New Brunswick a great place to live. We shifted the mindset of politicians and others alike. We worked together as an immigrant community to develop the province. We created systems to help new immigrants find work that matched their skills. We fought to be included in conversations about the future of work. We called our strategy "no decision about us without us." We fostered mutually beneficial relationships. We drafted documents for the education system. We helped international students succeed in changing how the province works.

Some people forgot their stereotypes of us. We did not bring in crime and terror. We did not create an economic burden. We provided community education, and we helped people to see the treasure that is New Brunswick. We brought so much to the province. We found jobs. Our presence brought more support to the province through federal transfer payments. We filled worker shortages that would otherwise have destabilized the economy and worked to support ageing populations. Our presence increased the ratio of workers to retirees, and we bolstered pension and healthcare funds. Our children, high achievers and upwardly mobile, promised future benefits to not only their families but also the provincial economy.

We encouraged immigrants to purchase homes, to create companies, and to invest in themselves. We developed new products, new production technologies, and new export markets. We worked in private businesses, hospitals, banks, construction, agriculture, forestry, and home building. We made business cases to expand and invest in the province. We formulated plans to attract new jobs and generate millions in tax revenue. We worked with the province to find ways to keep high school, college, and university students in the province. More and more people came to New Brunswick resolving the workforce challenge of not enough young people to replace retiring people.

We helped the economy grow at such a rapid pace that provincial and local governments were able to lower taxes and still have enough to pay for high-quality public services. Consumers received rebates that helped boost spending. The minimum wage increased to the highest rate in Atlantic Canada.

The increasing presence of immigrants improved the information and communications technology sectors. We supported small businesses and export-import industries. Our companies are competing globally. We helped create steady, good-paying jobs, which meant New Brunswickers did not need to leave to find work and raise families in other parts of Canada.

We achieved such successes because we were self-motivated and not afraid to retrain, to network, to adapt, and to become effective communicators. We worked hard and helped New Brunswick become the economic success it is today and will continue to be for years to come.

Part Four
Building New Foundations

Part Four: Building New Foundations

A Guaranteed Annual Income
Amy Floyd

I sit in my garden this weekday morning listening to the pileated woodpeckers who love living in the Nashwaak Valley as much as I do and remember the times in my life when I couldn't have a morning or a weekday off. In contrast to this gorgeous and worry-free morning, I recall times when I would spend days wondering where my next contract would come from, what I would be doing in a year, if I had enough to live on, or if I would have to file for bankruptcy on my student loans.

All my life all I'd known were "hard times in the Maritimes." Sometimes change, when it finally comes, comes fast. In the early 2020s, we had a pandemic and an election. A large number of citizens took their hopes and fears around climate change, the economy, and our social system to the polls and voted for a Green New Deal.

Nobody had all the answers, but we had to try; there were no other options. At first, many people didn't understand that just setting up more solar panels and creating technology to suck carbon out of the air would not solve the climate crisis. But then it became obvious we had created a global economic system that damaged the planet and people. We had to heal our relationship with the planet and with each other, if we were going to have a shot at survival.

After years of discussion and pilot projects, a big step was taken in a positive direction. Strong public support made a Guaranteed Annual Income possible. As far back as 1970, the idea had been tried in the little community of Dauphin in Manitoba. The project, called Mincome, or Minimum Income, had positive results but was cancelled after a change in government. But the idea persisted.

Mincome gave people in need a basic yearly income without having to apply for services or meet eligibility requirements or wait weeks for payments. Mincome gave up on policing the poor. The new federal Guaranteed Annual Income, as part of the Green New Deal, did the same.

There were a few false starts. In response to the pandemic of 2020, the federal government created Canadian Emergency Response Benefit (CERB), an effort to save the economy but not the people. Many of us were accused of defrauding the system. Some people would have to pay a good portion of the money back when tax time came.

It's no wonder people didn't want to return to work. A cheque for $2,000 every month was a dream, a respite and chance to catch one's breath. It might not seem like a lot of money, but for those working for retail giants on their feet all day, taking complaints, working five-hour shifts and making $1,400 a month, it was really something.

The Guaranteed Annual Income started with $17,000 a year. Not much money but a big jump from social assistance, then only $6,400 a year in New Brunswick for a single person. Anyone earning more than $25,000 a year would lose $0.50 on the dollar of the subsidy. The result was to get everyone in the lowest income brackets up to an income of between $17,000 and $34,500.

Workers who previously administered the social assistance, employment insurance, and disability pensions systems—none of which were any longer needed—were given the opportunity to do the minimal administrative work for the new income support system or to re-train for jobs in various sectors of the economy like human services and clean energy.

The concept of a Guaranteed Annual Income was at first daunting, even for people who thought it was a good idea. You see, we had been raised to think neoliberal capitalism was the only real option and we could hardly even imagine alternatives. The naysayers told us we could never afford it. It was too much money. But when the huge costs of administering and policing the old social support system disappeared and government stopped subsidizing the oil and gas industry, we were able to make it work.

The results were positive. Some people didn't want to work much, and some were unable to work. Overall, however, most people with low incomes worked just as much as ever but were able to stabilize their lives. With less stress, adults and children in families were both physically and mentally healthier. People were able to dream and plan and choose their futures. They could be more selective about the kinds of jobs and salaries they took on.

Kids knew they could go to a summer day camp and not have to worry about being chosen from a low-income waitlist. Parents knew they could buy good food and pay their bills on time. Seniors were able to get their homes repaired and hire more help when they needed it.

People could afford to try out a business idea or learn a new skill and still pay their rent and eat.

For me, I still work a lot. Every year, my networks increase, my earnings go up, and I don't need the Guaranteed Annual Income anymore. Even so, I can rest easy knowing that a few months in between contracts will not deplete my savings and I won't have to wait six weeks to get help.

The Guaranteed Annual Income helped me to plan and save enough to buy a bit of land and start a community permaculture education centre. We teach people how to grow food, build strong social systems, heal the land, and help reverse the effects of climate change. I don't know where I would be now if I had not had access to the Guaranteed Annual Income. I am thankful for it. My life today is better. The Guaranteed Annual Income changed our fears and anxieties into hopes and real, workable solutions that have made a stronger and more secure future for everyone.

Letters from the Future

Rural Assemblies
Abram Lutes

Farmers and workers, fishers and loggers, small producers, and rural people from across the province have gathered in Campbellton for the tenth biennial New Brunswick Grand Rural Assembly.

Crowds fill the Riverside Park as delegates from dozens of local Assemblies from both French and English communities mill around. We've organized Assemblies over the last two decades to respond to dramatic changes in New Brunswick. The banners are colourful, all purples and greens and golds and blues. The labour unions, the farmers' unions, the fishers' unions, the logging cooperatives, and the neighbourhood organizations are out in force.

It's hard to imagine that in years past Campbellton was known for economic decline, hard times, and a high suicide rate. Now, this town near the Quebec border is a model for rural economic and social development. The Campbellton region is remarkable for its food self-sufficiency, famous for its excellent health and senior care, its high levels of literacy, and the quality of its education system.

Campbellton is one of many towns in New Brunswick and across the country experiencing a social renaissance over the last few decades. Years of frustration with top-down bureaucrats, with centralization to urban areas, with civil-society groups operating out of faraway cities, and with environmentally destructive resource extraction, all boiled over into a grassroots, popular, and democratic movement in rural New Brunswick.

At first, these were spontaneous protests that had little to do with big ideas like "sustainability" and "rural revitalization." The protests were focused on local issues. Later, as the focus expanded and the language used expressed ideas about intersectional analysis, rural people began to draw connections between rural and urban poverty, between racialized violence in the cities and environmental violence and colonialism in rural areas. Frustrated by the "wait-and-see" promises of parliamentary parties on the left, right, and centre, people took things into their own hands. Rural Assemblies were central to this change.

Rural Assemblies first began not in New Brunswick but with other social movements in other places. There was the anti-austerity farmers movement in Greece, the *Gilets Jaunes* in France, the Landless Workers Movement in Brazil, and anti-racist and solidarity economics initiatives like Cooperation Jackson in the US. Like the New Brunswick's Rural Assemblies, each attracted comparisons to older forms of local democracy.

Some made comparisons to the local town meetings of New England, where farmers and townsfolk deliberated and found a consensus on local governance. Others fought for the Athenian forum, where the citizen class as a whole would govern the city. The best example was the comparison to Tommy Douglas's Co-operative Commonwealth Federation (CCF), which in the last century shook up the wheat farms of Saskatchewan and helped bring universal health care to all Canadians.

The truth is the Rural Assemblies were all this, and more; they were a political and organizational base arising out of conditions facing rural people in the twenty-first century. The climate catastrophe was worsening. Decades of austerity and free trade agreements had left the province impoverished. Indifferent governments were committed to urban-focused visions of progress.

The Rural Assemblies brought together workers, farmers, small producers, young people, unemployed people, and pensioners to craft demands and policies that spoke to the needs and desires of small towns like Woodstock, Belledune, Campbellton and Edmundston, St. Andrews and St. Stephen, and all the other the rural regions of the province.

The model is simple. Within each community, citizens assemble in a public space—a park, a civic centre, an arena. They organize community action and demand that governments pay attention. People form assemblies even in rural places without a central hub. Coordinating Committees are formed to implement the Assemblies decisions between gatherings.

As the model spread across the country, New Brunswickers took the initiative and hosted the first Grand Rural Assembly, which brought together social movements from across the province, the Maritimes, and the whole of Canada to share strategies, develop techniques, and to build campaigns.

The rural assemblies organized protests, and in one memorable action they demanded unilateral withdrawal from the disastrous Free Trade Agreement with the United States, and the creation of a Farming Board. The Rural Assemblies blocked major highways with manure and trees — a tactic learned from their French and Greek comrades resisting austerity in the European Union. They organized farming communities, drawing on farming models developed by early Acadian-Mi'kmaq settlements. Farmland was no longer managed as individual private property, and productive plots became part of an ecological and communal whole, redistributed annually based on the needs of the community and consensus among farmers. Assemblies used digital media and robotics alongside traditional knowledge and permaculture.

The Rural Assemblies also took initiatives to ensure that social services were available and appropriate for rural areas. They organized teams of community volunteers to pave roads and transport the elderly and impaired to their doctors' appointments, the basis of what would become the rural community public works program. They invited researchers and students to come stay in small towns and on community farms to exchange expertise, a project that evolved into the rural practicum program that became part of the curriculum at the University of New Brunswick, Saint Thomas University, and Mount Allison University.

Uncomfortable with local democracy, politicians first derided Rural Assemblies as "Hillbilly Red Neck fanatics tinged with old-fashioned, uneconomic, unrealistic, and anti-development ideas." But eventually the government had to concede the Rural Assemblies were a legitimate, representative force.

Federal and Provincial Commissions were struck and became liaisons between policymakers and the assemblies. The assemblies precipitated a sea change in New Brunswick and Canadian politics, not only in policy, which shifted from neoliberalism and free trade, towards public-led development and mutual assistance, but also in the very way political decisions were made.

Inspired by New Brunswick's Rural Assemblies, student, worker, and tenant associations in cities across the country formed urban region Citizen Assemblies. Popular will was not tempered by intermediaries like political parties and appointed experts. Instead, these democratic bodies were directly responsive to what emerged from the process of their meetings for initiating positive changes in public policy.

Part Four: Building New Foundations

It is at this Grand Rural Assembly in Campbellton, which includes representatives from Indigenous nations, where we are proposing a "good living" clause in the federal constitution that guarantees every Canadian the right to clean air, water, and a healthy environment. Emilie, a young woman from the north shore, read the proposal to the Assembly. She is one of representatives of thousands of workers and farmers and rural people in general who now use New Brunswick Rural Assemblies to speak for themselves and their communities in the creation of national policy.

Letters from the Future

Organizing Food Security
Louise Livingstone

As a long-time visitor to New Brunswick, it is incredible to see the change in the province's food system that has been accomplished. On visits to Fredericton we always loved to visit Boyce Farmers Market on Saturday mornings. We loved the variety of local meats, cheeses, fruits and vegetables, and talking to the producers. This feeling has now expanded to the whole province. There is a strong sense of belonging, and people are taking responsibility for what they eat. A whole new attitude toward food exists.

Eating in season is back, with people storing food for the winter based on their local growing cycles. People process summer produce for winter use. Solar dehydrators are in common use for drying tomatoes. People have rediscovered skills like growing and milling grains and making bread. Conifer plantations are being returned to mixed Acadian hardwood forests, which means a greater variety of wild foods are becoming available. First Nations have shared much of their traditional knowledge about foraging. Immigrants bring and share their own food traditions.

Working with many different groups and communities, the provincial government has agreed on a food policy that supports local agriculture and reduces food miles from farm to table. New Brunswick is becoming increasingly self-provisioning.

The local dairy industry is flourishing. The cows go out to pasture in the summer rather than being held in confinement all the time and fed a grain based diet. Artisans in commercial scale, community kitchens are making cheese from cow's milk, sheep's milk, and goat's milk. New Brunswick's fame for artisan cheeses is growing.

People now grow vegetables in their yards and community gardens. Some community gardens are like European allotments, with individual gardeners having their own plots. Others work collectively as community or neighbourhood. People who bring compostable kitchen waste to the community garden compost heaps are credited for receiving vegetables in return. Gardeners help others who have ground available but are unable garden. People save many seeds and thus maintain seed diversity and local supplies. Seniors, who were once gardeners, are valued in the new food system as they have many skills to share.

It's exciting to see landscape planning now includes vegetable gardens for new building where space permits. Buildings in high-density areas feature roof gardens, which helps reduce their carbon footprint and make them energy neutral.

All communities now compost garden and food waste. Food banks have developed into community food hubs. Food waste has been cut to a minimum now that stores are required to give unsold produce to the local food hubs where volunteers make meals. More and more grocery stores have gardens and growing systems that grow greens year-round.

My grandson, in grade ten, loves farming. He gains practical experience with growing vegetables and caring for laying hens at school. Students get a cooked breakfast and lunch at school each day. Senior students help prepare meals using the vegetables they have grown in the school gardens.

Schools now have tree nurseries. Kindergarten students plant tree seeds. By the time they graduate from elementary school, they have created new woodlands. Growing food and planting trees are now part of the province-wide land-based school curriculum.

How things have changed! New Brunswick has become a province of local food for local people.

Ending Homelessness
Nigam Khanal

I write this letter on a warm September morning from my cozy, beautiful home in Kathmandu. My house looks out from amidst the city at the beautiful mountains. This morning I received an email from an old friend in New Brunswick. My friend wrote: "Good News! The province of New Brunswick announces the end of homelessness." I read it in detail. Nothing could have made me happier today.

I was a young woman in my mid-twenties when I first landed in Fredericton in 2012 as an international student to join the Master's in Philosophy programme at the University of New Brunswick. I was passionate and full of dreams about my future in Canada. Fredericton seemed a beautiful city, especially in the fall when the trees turned different shades of yellow, red, and brown.

It was quiet and lovely, a perfect place to raise a family and to be a student — until I saw the dark side of the city. The homelessness was noticeable. In the winters, people were on the street in minus 25-degree temperatures, freezing and begging for money. It was painful for me to see them, especially when it was so cold that I could not imagine being outside for longer than half an hour. These people spent their days and nights outside in the cold.

The few homeless shelters in Fredericton were overfull with no capacity to take in more people. I volunteered for the community soup kitchen where people could come for a meal twice a day. The number of homeless people increased, but the shelters did not increase their capacity.

My heart broke during the 2020 COVID-19 pandemic. When buildings and services had to be shut down for more than three months to control the spread of the virus, vulnerable populations suffered. Some homeless people were living in tents in the heart of the city.

Reading my friend's news that homelessness has ended in New Brunswick, I feel glad that the province that was my home for more than eight years has marched ahead to show the world how socialism is possible in the first world.

I had first felt a gradual shift back in the early 2020s, when people were paying attention to racial injustice. Amidst COVID-19 and a recession, the Black Lives Matter movement in the US and Canada mobilized communities in the province. Soon after the 2020 provincial election, people started to organize under a new political party. After the hollow promises of the mainstream parties, New Brunswickers were finally waking up.

Eventually, New Brunswickers chose the honest leaders who knew how to put theory into practise. The new party with its socialist orientation won an election, and changes began to happen in the province. Ending the legal loophole that allowed the large oil monopoly to account for its profits in an offshore tax haven was an immediate priority. This required strong pressure from the provincial government at the federal level, but the change was made. The tax avoidance of the oil monopoly came to an end. Other New Brunswick businesses that had used offshore tax havens now had to pay taxes in Canada as well, and a large stream of new revenue became available to support social programmes in the province.

The province invested in housing for all, education, and eco-friendly industries. The provincial population has also been growing, as people from all over the country move to New Brunswick for better opportunities. The government followed through with its election promise to end poverty in the province through a Guaranteed Annual Income and to guarantee housing for every family and individual no matter their income level.

In a few short years, the new provincial government has been able to achieve the changes it had promised. Homelessness has been eradicated and no one is freezing on the streets. I am so proud of New Brunswick for its massive turnaround. It is now the first socialist province in Canada.

Letters from the Future

Transforming the Nature of Work
Christine Wu

Here in the wondrous quiet of my annual writing retreat in the woods, I have been reflecting on the nature of work — what it is and why we do it. I vividly remember how it felt to wrestle with the incessant hold and resulting fatigue of daily work up against the need to survive. Then, the global pandemic of 2020 hit, upending all established norms of workaholism, productivity, and the illusion of economic growth.

Like many others in New Brunswick, I was laid off when COVID-19 hit and began relying on the Canada Emergency Response Benefit (CERB) to pay the bills. Governments struggled to keep the economy afloat while the pandemic continued to destroy the status quo. CERB evolved into the Canada Recovery Benefit (CRB) as an attempt to tide things over until "normal" could be re-established.

What these benefits did not address, however, was the identity crisis that took hold of the many workers whose daily lives had been completely transformed — laid off, furloughed, or haphazardly thrown into the front lines. While front line workers wrestled with the ethics of working through a pandemic with little preparation or adequate compensation, questioning and pushing back on it, others were forced to reckon with their acute loss of identity. Without work, who were we?

At first, many resisted, filling all the corners of this unfamiliar new time with activities: copious bread baking, podcasting, binge-watching reality television. Unsurprisingly, these new rhythms were unsustainable. There's only so much we could do to keep the existential demons at bay.

Not only did COVID-19 force us to reckon with where we found meaning and identity, as death tolls rose around the world, it also forced us to confront our precarious mortality. We realized there was more to life than work. We wrestled with the age-old question about the meaning of life.

We needed new individual and collective identities that weren't dependent on the myth of economic growth and unending consumerism.

As life slowed down, it seemed we finally began to rediscover the inherent beauty in the world around us, away from capitalism and exploitative resource extraction. This slowing down forced us to look inward, away from finding our value in work and productivity and, instead, attend to the humans—and the more-than-human—that brought meaning into our lives.

When 2021 rolled around, with no clear end to the pandemic in sight and after having experienced a significant shift in daily life and societal values, the Canadian government transitioned CRB into an Guaranteed Annual Income (GAI) by using funds collected from finally implementing tax reform. Naysayers argued GAI would turn the nation into a lazy, welfare state. But the results astonished everyone.

Freed from the paralyzing oppression of the 40-plus hour work week and the Protestant work ethic, people didn't work less, they just worked differently. There was finally time to breathe, to look inward, to re-assess and re-discover what gave us meaning. Many of us found joy in our pandemic hobbies. We continued gardening and bread baking, shifting further away from reliance on corporate farming and industrial food-ways. Others went for long walks in the woods, attending to the long-forgotten wisdom in the presence of trees. Still others made art that not only created personal meaning but also spoke back against centuries of systemic oppression. By attending to ourselves, and the world around us, we inevitably made space for the earth, and ourselves, to heal.

With this new shift in values, governments decentralized, allowing for a more nuanced attending to various community needs. What surprised us all was the ripple effect; decentralization affected not only government structures but also our very being and living. Humans finally started to see that the world does not centre around *us*.

Ursula K. Le Guin, novelist and poet, once said, "We live in capitalism. Its power seems inescapable. So did the divine right of kings. Any human power can be resisted and changed. Resistance and change often begin in art, and very often in our art, the art of words."* At the time, these words were both powerful and hopeful, yet it was often hard to envision such

---

* Ursula K. LeGuin acceptance speech. *The National Book Foundation Medal for Distinguished Contribution to American Letters.* November 19, 2014. Copyright Ursula K. Le Guin, https://www.ursulakleguin.com/nbf-medal.

significant large-scale change. We did not foresee the catalyzing effects of a global pandemic and the sheer tenacity of the human spirit to find its way back to that which is meaningful and beautiful.

Le Guin was right. It was the artists and dreamers who helped tip the scales away from a crumbling capitalist society. With our faith in a corporate-run capitalist society fully shattered, we turned to and learned from those on the fringes of society who were now inspiring the world with their eye for beauty, for play, for conservation, and for meaning-making. By shifting our focus away from meaningless productivity and the dogma of work, we began to truly *see*. A desire and capacity grew to care for and connect and reconnect with our communities, our environment, and with ourselves.

Some people call our current time the post-work era. To me, that's a misnomer. We haven't stopped working, we are simply re-visioning work, discovering or rediscovering the joy of attending to ourselves and those around us in this new way of life devoted to meaningful work.

Part Four: Building New Foundations

Education First: Everything Else Follows
Naomi Gullison

Education matters, especially in New Brunswick. Now, as in the past, so much of the work done to maintain our society, and more and more of the jobs, require some form of post-secondary education.

Many years ago, the cost of obtaining a post-secondary degree was far too high. Young people were forced to take on gargantuan debt just to get an education, and, even so, the chances of landing a reasonable job were often discouraging.

Education at all levels suffered from the prohibitive cost of post-secondary studies. Universities were so expensive it became difficult to preserve bilingualism in New Brunswick. Now, our well-educated, debt-free, bilingual citizens are invested in enriching the province.

New Brunswick once struggled to find teachers able to work in both French and English. Math teachers were also hard to find. Now, there are plenty of bilingual language and math teachers in every community in New Brunswick, large and small, all working hard to pass on their knowledge to the next generation. The challenge was finding teachers fluent enough in French to teach immersion. The first step to solving this problem was to rethink how the province invested in *post*-secondary education.

With all that followed the response to the killing of Chantel Moore and Rodney Levi by police in New Brunswick, residents in various jurisdictions around the province began to realize that situations of social and mental health crisis should not be dealt with by the police; they should dealt with by social and mental professionals trained in crisis intervention. The biggest problem was funding the training programs for these professionals. This was addressed in part by refocusing police work on its core mission and shifting the part of its funding thus made available to help support education and training in the field of social care work.

This highlighted the whole problem of inadequate funding for education in general whether in language teaching, community care, public health, or skilled trades; there were not enough qualified people to get the jobs done. Attempts at structural reform proved ineffective. Finally, it was realized that only by investing in post-secondary education—in colleges and universities—that enough skilled professionals would be trained to meet the needs of New Brunswick across the whole range of essential services and development opportunities. Once this understanding became widespread, the barriers to investment in education were removed and New Brunswick was able to enter a new era of job creation and economic growth.

What did the investment in education look like? At first, it took a strictly needs-based approach. But this created unnecessary and costly amounts of red tape, excluded many students in need, and created conflict between students who received funding and those who did not. The basic idea of the program was that if a student's parents or guardian made under a certain amount of money each year, their education was free.

Advocates told critics that while some students were at first excluded from these grants, the goal was to keep raising the qualifying income until virtually all students were included. This plan ignored the reality that incoming conservative governments gutted the programme. This needs-based policy was less about achieving accessibility to education, than it was about the comfort of politicians seeking praise for passing it.

The alternative the province eventually pursued was to undo the neoliberal tax cuts of the 1970s and 1980s, and aggressively tax the companies that owned much of the productive forestland in the province.

"But oh," the critics cried out, "what if they leave? What if they pack up their suitcases of money and leave New Brunswick once they have to pay their fair share of taxes?"

But they couldn't. The old wealth of the land barons that controlled the province is based on raw material extraction. For them to leave the province, to give up their wealth in land and logging rights because of higher taxes would be suicide. So they stayed and they paid the taxes. When we taxed the wealthy and the corporate owners of New Brunswick's land and logging rights at appropriate levels, increased public revenue became available for investing in education.

We overhauled the universities. At first, tuition was frozen at all New Brunswick post-secondary public institutions. Then, tuition costs were steadily reduced to a reasonable level.

What followed was an economic and cultural renaissance in the province. As the level of education rose, people found more and more opportunities to escape poverty through both social programming and access to new jobs. And all of this was enabled through education.

How did we get there? In part, it was an active and vibrant student movement that emerged. It was the students who organized, who formed new unions, and who advocated for structural reforms and for fair taxes on the old money. The student organizers prevailed in solidarity over careerism and privatization. We created a vision of social and economic renewal and dragged the politicians along. They stopped prioritizing austerity budgets over needs of people and turned to making education accessible. And we paid for it all by the increase in tax revenue from the lumber and property barons. We have created a well-funded higher education system that is *reasonably* priced for everyone, and, hopefully, someday will be *free* for everyone.

The student movement that made such strides now stands in solidarity with the whole of New Brunswick's working class. The future of New Brunswick will be with advocacy groups taking a coordinated and multi-faceted approach to reform, rather than competing with each other and accepting the false narrative that only so much wealth is available to invest in the future of all New Brunswickers.

The future of New Brunswick will flourish when political leaders recognize that budgets are about investment not scarcity. Our future will be in recognizing that budgets, on a fundamental level, are about priorities, and that the intersection of all the struggles in the province is about making the people of New Brunswick and their access to education, a priority.

Letters from the Future

Restorative Justice: The Way Forward
LA Henry

The criminal justice system in New Brunswick is on the brink of embracing the restorative justice model as part of the alternative measures arsenal for offenders. Restorative justice offers a way forward — particularly for young offenders — that not only provides an opportunity for them to take responsibility for their behaviour but also to actually heal the harm done.

Our criminal justice system once lacked any significant recognition of the role trauma plays for both victims and offenders. Not all offenders are survivors of childhood trauma, abuse, mental health issues and/or addictions, but most are. And often, systemic racism, homophobia or transphobia are a part of this picture. Does this excuse harmful, criminal, or anti-social behaviour? No, of course not. The Criminal Code of Canada is clear that its primary objective in sentencing is deterrence and denunciation, with rehabilitation a distant third. But how are people actually deterred? How is crime actually denounced? What is the road to genuine rehabilitation?

When someone in New Brunswick is charged with a crime under the Criminal Code of Canada, they are arrested. Arrests most often involve handcuffs, with hands behind a person's back. This is the case whether or not they are resistant and whether or not they demonstrate any threatening behaviour. The restraints do not take into consideration age, race, gender, or the potential for trauma. If the arrested individual is detained in a detention facility, they are transported using shackles. These are leg irons that permit them to only shuffle. At a bail hearing, they typically wear leg irons and handcuffs (hands in front), and in court they are seated in the prisoner's docket. All of this occurs while they are still presumed to be innocent,

innocent under both common law and the Canadian Charter of Rights and Freedoms.

Who is arrested in New Brunswick? The dockets are full of the names of Indigenous persons and almost all folks on the dockets are living in poor conditions. To the credit of the administrative heads of the New Brunswick Legal Aid Services Commission, lawyers are available for all youth, but they do not get lawyers until after they have been charged and have attended their first court appearance. Due to government budget restrictions, adults charged with "summary offences" (minor offences) do not qualify for Legal Aid lawyers, unless they are likely to do jail time. No matter how poor, how illiterate, how challenged by disability or addiction, people charged with summary offences are expected to self-represent.

I have represented many first-time offenders. They are always surprised because they do not see themselves as *criminals*. It is often an *ah-hah* moment when someone realizes they are now being perceived in the same way they had themselves previously judged those who ran afoul of the law.

The criminal justice system treats people as though the lines of bad guys and good guys were as clear-cut as an episode of *Law and Order*. The line is not so clear-cut. It never has been. People are complex. People act out of complicated motives, which are often impulsive and unwise. Whether or not they are policed and whether or not they are charged is too often a matter of life circumstance beyond their control. Arrests, detentions, and court appearances are themselves crisis inducing and traumatic for many, if not most, offenders. These practices, once designed to function as a form of social shaming, have continued to operate as such.

But now, after a long battle, the Province of New Brunswick is beginning to invest in the use of restorative justice as part of the alternative measures and diversions available in its criminal justice system. This is good news — very good news! While both adults and youth can be offered alternative measures, the *Youth Criminal Justice Act* now requires that such measures be considered before criminalization occurs. But, until recently, the practice of policing in New Brunswick did not see nearly enough youth go through diversion.

Indeed, we now see that all youth should be given the opportunity to be diverted from the court system. Alternative measures provide a wrap-around service that seeks to address the youth's needs underlying the criminal behaviour. It seeks to understand the behaviour from a youth-centred focus. It seeks genuine rehabilitation.

Restorative justice circles go a step further; they provide the offender with the opportunity to listen to the victim and hear the harm that they have done. It permits the victim to have a voice in directly confronting the offender. It provides an opportunity for community members who have been impacted by the offensive behaviour to participate. It provides a possibility for genuine healing.

Thank you, New Brunswick, for stepping up and seeing that this is a way forward that will decrease the trauma of the criminal justice system especially on youth and provide the way for many first offenders to a positive and hopeful future.

Part Five

A Resilient, Low-Carbon Way of Life

Part Five: A Resilient, Low-Carbon Way of Life

New Brunswick's Green New Deal
Daniel Tubb

The New Brunswick Green New Deal began to seem possible about a decade ago, when spring floodwaters ravaged New Brunswick, the Ottawa River washed through Ottawa, a dike levy broke to waterlog Montreal, and the list of small towns and villages flooded across Ontario and Quebec seemed endless.

The consequences of climate change weren't hitting only Eastern Canada, though. During those hot summers of the 2020s, British Columbia and Alberta experienced droughts with vast wildfires sweeping their interiors. The blazes during those years made the fires of 2017 and 2018 seem like small dress rehearsals. Climate refugees swamped Vancouver and Calgary. At times, cities seemed like war zones, but the bombs were climatic. The country was not ready for the devastation, as we were still dealing with the economic, social, and health turmoil brought on by the global pandemic.

To many Canadians, global health and environmental catastrophe had previously seemed so far away: mudslides and wildfires in California, hurricanes in Puerto Rico, floods in Houston, cyclones in India, or the HIV/Aids pandemic in Africa. But early in the 2020s it became clear we were all connected. It wasn't the global pandemic that changed the world as we knew it; nor was it just climate change—it was the climate catastrophe.

The catastrophe was undeniable, and yet governments and corporations were in denial. They were too preoccupied with the economic turmoil and a public health disaster. We were already living with epochal change, and we were not doing enough. With the scars of fire, flood, and pandemic on our minds, we began to wonder, "Is there nothing we can do?"

The answer was clear; something could be done. We realized the old economics was not only bankrupt but also had caused the problems in the first place. The economists had missed the obvious; economic growth on a finite planet leads only to disaster. The arithmetic was the simple one that young families crowded into condos with two kids during a lockdown understood. There wasn't enough space for everyone on the planet to live like people in Canada did—we would need 4.7 planets.

It was a time of not only realization but also action. People began to mobilize and organize. Students held regular climate strikes. The environmental rebellion came to North America. High school kids from New Brunswick crossed the country by train to galvanize support. Black Lives Matter took to the streets to protest police violence and demand an end to white supremacy.

In a country shaken by illness, by floods, and often burning, those years were when students, First Nations communities, unions, workers, Black activists, environmentalists, and many voters began to see that doing nothing in the face of climate catastrophe, a pandemic, and the legacies of white supremacy was not inevitable, it was suicidal. Alternatives existed.

There were elections. The conservative parties split the vote on the right, and the scandal-riven Liberals fell back on their clichés and tired old neoliberalism tropes from the previous century, as bankrupt as Communism had been before the fall of the Berlin Wall. The market will save us, was all the Liberals seemed to offer. Their government's economic support for the pandemic did not go far enough. Their carbon tax was far too little, far too late. Their taxes shifted the burden of climate change onto the poorest and least able to pay. The Liberal government, intensely unpopular, fell. Not to the Conservatives, of course. How could it? The Conservatives were still denying climate change, their leaders strategizing to save the oil companies in Alberta. Instead, the Liberals became a minor partner in a Grand Green Coalition for Climate Change.

The Green wave went very far, indeed. The Coalition swept to power on the promise of a Green New Deal. Over the next ten years, Canada implemented an aggressive plan of climate transformation and decarbonization, of investments in public health, in energy, infrastructure, and efficiency, in taxing and shutting down carbon emitters, in restarting the economy for millions with new training and creating new green jobs of retrofitting inefficient buildings, and in transforming the popular financial support for those who lost their jobs to COVID-19 into a Guaranteed Annual Income which helped win support and transform an economy away from dirty carbon. Together, it was this shift that allowed us to surpass the Paris Climate agreement by 2030, and for our country to decarbonize.

It was exciting to be alive. We barely made it, of course. It was a hard decade, in some ways harder than the Great Depression and the Second World War. Certainly, the lingering health impacts of the pandemic were a bitter pill to swallow on top of such heartache. But we did it. I'm much older now, my son is 16, my parents in their 80s, and although few people thought we would achieve this goal, we are on track to limit climate warming to below two degrees. The path towards decarbonization and negative carbon emissions is the key to this accomplishment.

With the pandemic, the fires, and floods, something changed. It was a crucial time. Canadians not only saw the climate catastrophe wash into their homes and burn their communities, they also overcame their despair and began the hard work of changing the policies and practices that had brought on the first wave of climate catastrophe. Politically united around a Green New Deal, a majority of Canadians are now determined to help stop further climate and environmental deterioration.

Letters from the Future

And then Faster than We Thought Possible...
Erin Seatter

I was a new mother with a young child at home. I worried about the future, her future, but I was also busy caring for her, morning to evening. Every day we read children's books together. We went through ABC books about insects—ant, bumblebee, caterpillar—and animals in Africa—aardvark, baboon, cheetah—and animals in Canada—auk, beaver, caribou. We read about a panda whose sneezes upend everyone around, about a brown bear that sees a red bird that sees a yellow duck looking at them, about a gorilla intent on escaping its cage to sleep in the zookeeper's bed. We laughed at the pigs singing la la la, the turkey with a shoe perched absurdly on its head, the moose and the goose together having juice; all these creatures were happy, all were well.

Other families did the same. How many wondered which animals would disappear in their child's lifetime? How many worried they were teaching their children not so much about the wonders of the living world but an exhibit of soon-to-be dead artefacts? How many, as they smiled over the adorable white cubs climbing a hill of snow, considered the starvation facing real-world polar bears as Arctic ice succumbed to the onslaught of global warming?

We lived amid disconnection. We bought garments pieced together for pennies by women overseas working without breaks in unventilated rooms, laundered synthetic fabrics sending microplastics into the oceans, dined on meat supplied through the cruelty of industrial animal production and slaughter, and blocked human bodies from coming into Canada escaping violent regimes our country had equipped and supported. We couldn't reconcile *them* with *us*, couldn't see how we were all bound.

Part Five: A Resilient, Low-Carbon Way of Life

Climate catastrophes sped up. Temperatures rose, and, in the slow boil, far-right politics followed. Trump, Duterte, Bolsonaro, Modi, Erdoğan, Netanyahu. At home we saw white supremacists calling themselves white nationalists calling themselves citizens concerned about cultural protection and mass immigration. As the environmental and political situation worsened, it was hard to see what it would take to snap us out of our complacency.

And then SARS-CoV-2 crossed into the human world. This microscopic invader caused great upheaval as it crisscrossed the globe, bringing to a halt centuries-old activities said to be unstoppable. As cities locked down, industry halted. Smog cleared. In some places, including Kathmandu, long-obscured mountain peaks emerged into sight on the horizon. In others, residents inhaled clean air for the first time.

In the age of the new coronavirus, old fissures of class, race, and gender were laid bare, as was the estrangement from the natural world at the core of our economic system. Societies had to reconfigure. Business as usual meant death. In the uncertainty and havoc of the pandemic, solidarity became the only way forward. But even with that clarity, we still had to fight. The tyrants of the status quo did not bend easily. Remembering those who fought for us before we arrived, and those who would live with what we left behind, we rose up. We wanted possibilities other than needless death.

We came out against police violence. We put our bodies in the way of pipelines. We filled courtrooms with cases of civil disobedience, where everyday people explained there was no other choice, we had to change — the status quo meant a terrible future. Children argued they needed a world to live in. We made our banks, pension funds, universities, unions, and governments divest from the corporations killing the world. We lay down on roads and on airport runways. Students refused to go to school and went on mass strikes.

Then we elected people's governments, nationally and provincially. It took a lot of hard-working, honest people — people like us — to take back our political life from timid centrists and shipwrecking neoliberals. With a wildly popular tax on the super-rich, we set about making society work for everyone.

We rebuilt our neighbourhoods and cities and stitched them together with efficient rail systems. We placed solar panels on our roofs to harvest the sun. Alberta led the way, fulfilling its aspiration of becoming an energy leader, not with its now deposed fossil fuels, but by embracing its clean power potential. First Nations had to consent to projects on their land or projects didn't happen. They truly had the right to veto. But rarely did they exercise this power, for Alberta's innovations in the field of energy meant the end of tar sands poisons in their waters and pipelines through their territories.

Arctic policy shifted from Canadian claims to resources to justice for the Inuit, which meant preserving their environment and traditional means of subsistence. With careful fisheries management, ocean protection, and climate change control, marine life began to recover and flourished. The fishing sector was revived. New Brunswick became the first province to shut down the aquaculture that had spread disease among wild fish populations. Across the country, domestic recycling plants were established, so we no longer shipped our trash to Asia and Africa; we could deal with all kinds of materials here at home.

Instead of allowing urban sprawl to eat up natural habitats and rich farmland, we nurtured smart city planning and density development. We made wildlife movement corridors that were integrated with transportation routes. We at last realized we were sharing a home with millions of other creatures with a right to space and to life, and we stayed out of their way. We called an armistice to our war on the planet. Along the way, green jobs were created. As more and more people had decent, reliable, unionized employment, we were able to reimagine our economies so we could work less and spend more time with family and in pursuit of personal passions. We rediscovered community.

And then, faster than we thought possible, border walls disintegrated instead of ice sheets. Forests rose instead of temperatures. Storms and fires subsided, as did hate. Soon, creatures that had been vanishing—Acadian flycatchers, blue whales, cranes—refilled the space we left for them. Endangered ecosystems were changed, but they survived—as did the human species and our communities.

Part Five: A Resilient, Low-Carbon Way of Life

Watershed Councils Adapting to the Climate
Adje Prado

Over the years, the very meaning of governance in New Brunswick has changed.

The areas in New Brunswick previously known as Local Service Districts (LSDs) were transformed into Watershed Councils. Watersheds became an effective level of local government, both better empowered and better informed than LSDs. Residents are elected to Councils at the level of the watershed, and each Watershed Council has more local responsibilities, powers, and data for making decisions.

The Councils and their Councillors have the responsibility to manage the local economy, facilitate local development, provision local services, and support local ecosystem services, all focusing on the geographic area demarcated by the watershed they represent and to which they are responsible. Each Council balances its own economic budget and its water budget to maintain its constructed and natural assets in its territory. Their long-term planning has helped prevent flooding in lower lying areas of the province and have helped protect against drought in other areas as the weather became more extreme.

Taxes from at-risk areas of the province contribute financially to help re-engineer infrastructure and maintain natural areas upstream. The goal is to balance human needs and environmental needs. All watersheds are monitored using a mix of remote sensing technology and an extensive network of sampling and gauging stations. We developed this expertise in partnership with our research institutions, resource companies, and local communities after adopting widespread environmental reform, which forced innovation in the natural resource sector.

To assist this local planning, the Province has adopted an open-data initiative. The Province's environmental and social data are collected regularly and stored in a central database. The data allow different cutting-edge algorithms, researchers, and experts to compare and optimize at the most local level. The province has become recognized as a global leader in its approach to the integration data with the aim of social and environmental stewardship.

Today, the ecosystems across the province—the river basins, the wetlands, the highlands, the coasts, and of course the forests—are very different than before the peak of the global climate crisis. At the local level, we have adapted. Foresters and land managers have gone to extraordinary lengths to use adaptive silviculture and assisted migration to create new and more resilient versions of the Acadian Forest. They have drawn on ecological models, climate change simulations, and environmental data to make these changes.

On the coasts, in the highlands, and in the forests, a high-tech renewables sector has blossomed. The renewable industry has diversified its products and developed new technologies to process a more heterogeneous lumber supply. Carbon capture now accounts for a large portion of the forestry activities in the province. Some carbon producers pay to have quick-growing species cut and stored deep underground to reduce carbon dioxide levels to pre-industrial norms. Others pay to revitalize the richness and biodiversity of the new Acadian Forest. While advances in technology have facilitated better and more selective harvesting, clearcutting continues in some areas, to simulate natural disturbance and attempt to mitigate the forest fires accompanying the warming climate.

Aquatic ecosystems changed as invasive species displaced native ones. Still, a new balance has been achieved. From the warmed waters of the Bay of Fundy to the headwaters of the Wolastoq in the north, some species could not be preserved. But ecological functions have become far more central and careful planning and design has become the most important goal. Commercial exploitation has give way to sustaining the health and resilience of aquatic ecosystems as the overriding and central goal of fishery and oceans management.

In the face of the food scarcity resulting from extreme weather in California, Mexico, and other food producing regions, the province collaborated with researchers to develop agricultural crops adapted to the new local climate conditions. The goal was to support resilient, local agriculture. New Brunswick now produces much of its own food through permaculture and urban agriculture. Local food banks are nearly self-sustaining, thanks to the vertical farms they have established. This has been possible because of the low cost of renewable energy following reforms to the province's energy system. Rather than being a net importer of food, New Brunswick has become mostly self-provisioning.

To optimize resources and maintain services during the climate crisis, the province initially dissuaded population dispersal. This resulted in some smaller communities being reclaimed by nature, as their populations dwindled. Over time, New Brunswickers realized the problem with this approach, and saw the opportunities from attracting climate refugees and climate immigrants seeking alternatives to sweltering heat and extreme weather events.

Communities and businesses across the province lobbied the government to develop immigration programs and sustainable housing developments to provide the capacity to attract and accommodate new citizens in order to revitalize aging communities and solve labour shortages. Adapting to a new culture was difficult for some, but the province's approach — different from what other regions saw as a crisis — allowed immigrants to find suitable housing and learn not only at least one of the official settler languages but also one of the official Indigenous languages.

The growth in the population across the province brought a cultural and economic boom, turning New Brunswick's multicultural communities into tourist destinations and sparking innovation and entrepreneurship across the province. Professions that once were severely understaffed benefited from skilled workers arriving from other parts of Canada, North America, and abroad. Streamlined equivalency exams helped new arrivals enter their professions.

Most importantly, the education curriculum was revised to incorporate systems thinking into all courses. Younger generations began to wonder how it was that earlier generations made a distinction between the natural world and the social world and created a separation between nature and culture. How could nature have ever been an externality to the economy? Nature-based learning and outdoor classrooms have become the norm and children now understand their place in the natural environment.

With Watershed Council governance in place and the Provincial Government in full support of this regional and local empowerment, it is this new generation that will ensure the mistakes of the past are not repeated.

Letters from the Future

The Flip to Low-Carbon Mobility
Matthew Hayes

What a difference a decade can make! Canada had just elected a parliament no one was quite sure was going to do very much about climate change, let alone about public transit. But then events took their course.

Most middle-class people owned their own cars and drove to work. This is no longer the case.

For many decades, Fredericton's transit planning focused on sending buses across the city, attempting to cover as much space as possible. The goal of the 2008 Transit Plan was to ensure a high percentage of the city's residents were within 500 metres of a bus stop. What this meant was that the city's bus lines wriggled all over the city as development sprawled outward.

Moreover, it meant that taking the bus was not really practical for most people, and very few people took it, even if they could walk to a bus stop. Service was not frequent, and it was also not quick. To get to the Regent Street Mall, some lines would first go all the way out to Lincoln. All buses met on King Street downtown, where they idled for 15 minutes every hour, ruining the streetscape of one of Fredericton's main drags with toxic diesel fumes.

Because it was so impractical — there wasn't even service on Sunday — most people didn't rely on the bus service to get around. But those that did relied on it a lot, and it constantly failed them.

The coronavirus pandemic created a new fiscal environment in which the Bank of Canada met budget shortfalls and new investments. This continued throughout the 2020s, and this public financing technique is now increasingly used for public benefit, especially with respect to climate change mitigation. Therefore, municipalities have more federal funding to make more ambitious transit plans that reduce costs for Canadians.

First, Fredericton used new federal funding to extend public transit service to Sundays. But from there, it went much further to increasing the efficiency of urban mobility.

In 2020, Frederictonians were paying a lot to own and run cars. Factoring in depreciation of the asset, the cost was on average $9,000 a year. With 1.72 cars per household (according to the 2016 census) and 26,328 households, that amounts to $407.6 million per year.

Traditionally, municipal governments sought to increase the tax base through property development, and thereby also expand the cost of the city for individuals, since property development occurs around its edges. But the coronavirus crisis forced people to start getting more innovative, and to start thinking of ways that municipal policy might instead redirect individual spending to more efficient, shared public services that reduced the cost for everyone.

In 2019, Fredericton spent $10 million on its bus service. Part of these costs were recouped. In the 2020s, that number expanded, and today, we have much better urban mobility for a whole lot less than $400 million. If we wanted to, we could even invest in new infrastructure, like inter-city light rail services linking to Moncton and Saint John via Sussex, and a new walking bridge across the river on the Carleton Street piers.

Fredericton improved the reliability of its transit by designing new routes, like the fast transit route between the Brookside and Regent malls. Gone were the days of the 'hub-and-spokes' system centred on Kings Place.

Instead, the city brought in new planning regulations mandating development ratios and providing incentives for mixed-use density along transit routes. By 2030, a new high-density streetscape had emerged with retail space on Dundonald and Beaverbrook Streets, and on Smythe Street along the old Exhibition grounds, and on Main Street on the Northside.

This had several other desirable knock-on effects. It completely changed the University of New Brunswick and Saint Thomas University's place in the city, giving them a high-amenity, walkable neighbourhood that suddenly attracted students from around the country. Recruiting to UNB and STU in larger centres became a whole lot easier, boosting enrolment. That increased the number of young people in the city, and a large number of them began staying in the city year round, contributing to a vibrant music scene now rivalling the one in Halifax.

New provincial funding for an Age-Friendly Cities initiative has given new life to the city. Older residents, many of them living in isolated apartment buildings, now benefit from new funding to improve public transit infrastructure. This makes apartment living a lot more desirable, as well as affordable.

When you need a car, you can get one, but it is not like it used to be. People don't own their own cars anymore; they rent access to a car fleet managed by a non-profit public agency for the public benefit. The Canada Mortgage and Housing Corporation—yes, CMHC—teamed up with Canada Post—yes Canada Post—to design new, public ownership of automobile fleets. And when you have to "gas up", you do so by recharging the battery power, courtesy of NB Power charging stations.

The mobility sector, as it is now called, has created hundreds of new jobs in Fredericton, some of them with high tech firms that run infrastructure systems in other cities. These are not the old call-centre jobs. They are unionized jobs—some in the public sector—that pay highly qualified personnel a good wage with benefits.

The future of the city, it seems, is rising from its empty parking lots. Massive new housing developments are filling in the city's many parking lots that once spent most of the day empty and grey; they now teem with the vibrant and colourful life all the time.

Perhaps the most pleasant surprise of the last few years is how greater urban density in Fredericton has led to new types of sociability that differ entirely from the 2010s, when social media was all the craze. Today, socializing takes place way more often face to face. There are more people in Fredericton moving around. People run into one another more often.

We don't know yet whether the efforts we have made are enough. We are still decarbonizing our civilization at a global scale. The level of cooperation is unprecedented. Thankfully, we built on the experience of the coronavirus vaccination programme. Acceptance of a few sacrifices is now an accepted norm, replacing the cynicism so common in the 2010s. It has led to a more optimistic view of the future.

Part Five: A Resilient, Low-Carbon Way of Life

Continuing the Slow Decline, but that's Okay
Cheryl Johnson

New Brunswick has long been referred to as a "have-not" province because it has a history of receiving federal government transfer payments to help support provincial government services. In fact, most provinces receive equalization payments, so there are more 'have-not' than 'have' provinces.

New Brunswick has been considered not only a 'have-not' province but also a 'has-been' province. The province's heyday was around 1867. As one of the four founding provinces of Confederation, New Brunswick was an epicentre of global shipbuilding. A ready supply of lumber and year-round ice-free ports made the province a powerhouse. However, as more provinces joined Confederation, the centre of Canadian innovation and industry moved west.

With the benefit of hindsight, one can see how the economic prosperity of the province had been based on chance and an accident of geography. Many hard-working people had helped spur things along, but the 19th century shipbuilding boom was mostly good fortune tied to the province's forests and proximity to European ports. As the need for wooden ships disappeared, New Brunswick did not replace the industry, adapt or change to new technology. Instead, it began a long slow fade. The big timber had been cut and little came along to replace the industries tied to this resource. When the St. Lawrence Seaway opened in 1959 and brought ocean-going ships into Upper Canada, New Brunswick's maritime significance declined even more.

Much like the history of that twentieth century decline, the decades since 2020 have been more of the same for New Brunswick. Regions relying heavily on refinery oil and clear-cut forestry practices have experienced a tough go and the adjustments have been difficult. When the sun set on the petroleum industry, it was a quick death. With the need to reduce carbon in the atmosphere and move away from oil and gas, Alberta's oil sands became worthless. The Irving refinery in Saint John closed down. The industry that had once brought such good fortune to many in Alberta and to a few in New Brunswick, faded into irrelevance. The big change hit when a large part of the global community had had enough of our gas-guzzling, resource-stripping ways. Sanctions were threatened against Canada, companies were shut down, and those remaining had to pick up the pieces.

Innovation came, but it came from away. Our government and industries were far too slow to adapt. Cheaper and more efficient renewable energy and the creation of new battery systems and sources of power changed the way we produce, store, and consume energy. People no longer need fossil fuels. But the profits from buying and renting and licensing all that equipment went to the global conglomerates with few ties to Canada. We used to be an important economic powerhouse but not anymore. In the end, Canada was a petro-state too stubborn to invest in new technology and new industries. The batteries and the technology behind renewable energy were produced elsewhere. We have become an economic backwater.

Thank goodness we can benefit from the technologies developed elsewhere. If we had been waiting for the renewable revolution to occur here, it would have never happened. In New Brunswick, the slow decline continues. Our infrastructure, lacking adequate investment, continues to crumble. In that sense, it's not much different from decades ago. Now, as then, people by and large get by. The weather is worse, but not as bad as some places. We all relate to each other and we all commiserate about our shared experiences of a changing climate.

Maybe the slow decline is for the best. If New Brunswick were the centre of innovation and prosperity, we would not have the space and time to be ourselves and have the freedom of the wild woods. Now that the large companies and wealthy families have gone, there is less conspicuous consumption and more equality. It's far easier to live a small and sustainable life in a 'have-not' province in a 'has-been' country. Life isn't so bad, and we make do. In a way, we have come full circle.

Think of the pictures of old homesteads and farms in the early years of European settler colonialism. The lifestyle now, is more similar to what these pictures show than to the state of affairs in the early twenty-first century. The infamous "Age of Plastics" has passed and only the memories and the evidence in the landfills and oceans remain. Those of us who stayed in this place are in a new age of peace and hardship with the land. People are working together and sharing knowledge and compassion. We must be united, or life would be too hard otherwise.

Much work remains to be done, but the work is satisfying in a way that old corporate jobs of the past never were. Those of us still on this land are back where we truly belong—in the garden, forest, and field.

Letters from the Future

From Business as Usual to a New Prosperity
Carl Duivenvoorden

Early in the 21st century, it was business as usual in New Brunswick. Unfortunately, "as usual" came to mean "not very good." The world was changing greatly, but somehow much of that change bypassed us. Successive leaders stayed the course on New Brunswick's economic tradition of resource-based industries and megaprojects, and we bumped into some disconcerting tipping points.

Virtually all the big, high-value trees were gone from our woods. We struggled with the realization that our northern forests cannot compete with tropical plantations that grow fibre much faster than we could ever hope to, and we were left trying to figure out what comes next.

Our economy seemed to ebb and flow on the whims of megaprojects. Cash-strapped governments could not afford to be selective, and readily endorsed proposals like the Sisson Mine, which promised jobs in return for one-time extraction of minerals in the sensitive headwaters of the Nashwaak River. But in the end the mine was never built; the company behind it failed to attract investment. It eventually gave up, and our economic see-saw continued.

Shale gas tempted us with similar promises of wealth and employment, in spite of widespread misgivings about its role in causing climate change. Perhaps just as well, it too failed to materialize.

True, megaprojects of the past did bring good jobs, but most of those proved temporary. Our provincial resource cupboard became depleted, and investment money went elsewhere. Business as usual was not working.

Fortunately, at that low point, a new generation of leaders came on board and there's now a positive vibe in New Brunswick. They recognized the need to transform the province and build an economy based on clean

Part Five: A Resilient, Low-Carbon Way of Life

energy. Sure, there were challenges, as there are with any bold initiative. But our leaders persevered and now here we are, and it's amazing.

Locally assembled solar panels cover virtually every roof in the province. On a clear day, even in January, they provide much of the power we use. Homeowners love them because they turn sunshine into money. NB Power still generates electricity, but only from hydroelectric dams, wind turbines, and cutting-edge tidal power generators. The company's larger role has become that of power traffic director; its smart grid precisely balances power supply and demand, ensuring everyone gets what they need. The batteries of thousands of electric cars buffer that smart grid. Those batteries get charged when the sun shines, the wind blows, and the tide turns, but they in turn supply the grid during demand peaks. And as a bonus, our air is cleaner than ever.

New Brunswick's construction industry is thriving, thanks to a building code mandating that all new buildings be energy "super-efficient." Electric baseboard heaters are history. We now use solar hot air systems, super-efficient heat pumps, and locally grown biomass. Most New Brunswickers agree that sustainably harvested home grown energy is the best use for our forests.

Most importantly, a culture of efficiency has taken root. We've leveraged the traditional resourcefulness and ingenuity of the peoples of the province and developed an expertise that is the envy of neighbouring jurisdictions. We've become an incubator of ideas, apps, and patents, fuelled by the new sense of confidence and empowerment that has swept the province. We no longer send our dollars away in exchange for oil, so we have many more to spend here at home.

It's the confident, prosperous, sustainable New Brunswick I've dreamt of for my children.

Connected Communities
Susan O'Donnell

I'm old now but still remember the year when a road accident sparked massive positive changes in New Brunswick. The Progressive Conservative Premier and two cabinet ministers were returning to Fredericton after a breakfast briefing at Irving Oil headquarters in Saint John, when their speeding car hit a new flash flood washout caused by the clear cutting of a nearby forest hillside.

The three PC politicians were unhurt but, with a broken axle and no cell service, they missed the vote in the Legislature on the third reading of the Green Party bill to revise *The Electricity Act*. Without the three "no" votes they would have cast, the bill passed into law. The new *Electricity Act* allowed communities to develop their own renewable energy plugged into the NB Power grid. All fifteen First Nations took up the challenge, and within a few years, their wind, solar and geothermal power plants, connected to storage technology across the province, were a main source of energy for the NB Power grid.

NB Power's new, forward-thinking CEO bought more renewable energy from neighbouring provinces and was able to shut down all the dirty energy generating sources in its provincial system. First, NB Power closed its coal-fired plant in Belledune, then its heavy fuel oil plant at Coleson Cove. Then, it decommissioned its Lepreau nuclear plant, which took years with the nuclear industry's whining and foot-dragging and cost many times more than was reasonable. All the workers were unionized and retrained. Some chose jobs cleaning up the toxic soil in Belledune and dealing with the tons of radioactive waste at Point Lepreau.

Finally, NB Power partnered with the First Nations and renamed itself. NB Power became Wabanaki Energy and began a comprehensive grant and loan-payback program to retrofit homes and other buildings, making them more comfortable and less expensive to heat. Retrofitting New Brunswick created good green energy jobs in every rural and urban community.

Part Five: A Resilient, Low-Carbon Way of Life

With funds generated by their energy windfall, the New Brunswick First Nations took on the telecom giants. They hired the best litigation lawyers in the province, lawyers who had once honed their skills as goons on government contracts to fight public sector unions at the bargaining table.

The First Nations launched a strong legal case for Indigenous rights to the radio frequency spectrum for cell phone communications. Years before, the telecom giants had taken control of the spectrum without recognizing that the Treaties included the air above the unceded and unsurrendered Indigenous territories in New Brunswick. After years of litigation, the First Nations won their case for spectrum rights and were awarded billions of dollars in damages.

With their win and spectrum allocation, the First Nations created a new provincial telecom utility for cell phone and Internet service, Wabanaki Commons. Once the utility was running smoothly, First Nations gifted the ownership to everyone in New Brunswick in recognition of our shared treaty relationship. New Brunswick residents thus became members of the Wabanaki Commons cooperative, owning and continuing to invest in excellent Internet, videoconferencing and cell services, and creating small businesses to use the digital services.

Wabanaki Commons' extensive network support service trained local technicians living in every community ensuring permanent local jobs were created across the province. This attention to local needs guaranteed quality Internet, video, and cell services in all rural areas.

Rural communities across the province trained and empowered local elected representatives to work with other levels of government and Wabanaki Commons to re-open and build new community schools, virtual classrooms, health centres, telemedicine, courts of law and many other virtual public services, reducing the need to travel to access public services.

The result was a steady migration of professionals from urban to rural New Brunswick. They were able to work in beautiful rural environments and share their expertise with urban people using the Wabanaki Commons video network. Moncton, Fredericton, and Saint John began to lose their populations until immigration and internal migration from other parts of Canada balanced this out.

Well-established wind farms on the Acadian peninsula, offshore, and around the province were also thriving. Acadian community revitalization was well underway as young people stayed home to work in hundreds of family-run businesses. Home-based businesses used the Wabanaki Commons network to deliver French lessons by video and provide French-language business support services across Canada and around the world. In the summer, the communities welcomed language students of all ages into their homes for French immersion. The summer visitors worked alongside the Acadians on the community wind farms, trouble-shooting Internet problems, and helping with the ongoing building retrofits.

Eventually, the University of New Brunswick made the decision to close its Fredericton campus and be an entirely distributed educational environment. The departments relocated themselves across the province, interconnected with each other and with students by the Wabanaki Commons video network. They thrived in rural communities far from the control of UNB senior management.

The former UNB Fredericton buildings were retrofitted and renovated to make affordable housing units. Under the leadership of a well-established urban teaching farm in Fredericton, the residents and food security trainees turned the former university land and other green spaces into community food gardens. Community garden projects were also thriving in backyards and greenhouses across the city, supplying fresh produce for residents year-round.

In Ottawa, the Prime Minister, leader of an Indigenous Solidarity-Green-NDP coalition, launched the national electrified rail, energy transmission, and zettabyte fibre telecom network infrastructure across the country. Connecting the Port of Belledune with the Port of Vancouver created a multi-purpose energy, telecommunications and transportation corridor stretching from the Atlantic to the Pacific Ocean. Electric freight and passenger trains became the backbone of the country's public transportation system, with hubs and spokes plugged into the corridor to eventually connect regions and communities across the country.

Belledune became the Eastern terminus for the rapid electric trans-Canada passenger train (the Coast Salish-Wabanaki), and scenic Baie-des-Chaleurs became a national tourist destination. The Belledune beaches are packed every summer.

Many visitors to the area do the popular hike from Belledune to see the masses of purple flowering rhododendrons covering over the former lead and zinc smelter, phosphoric acid plant, and coal-fired energy plant, industrial dinosaurs once polluting the local environment. When the plants closed, the workers had their choice of green economy jobs. Most chose to train for opportunities in the many small businesses opening near the national rail corridor.

    I moved to Belledune and am writing this letter from here now, in my little piece of paradise—a retrofitted cottage overlooking the sea.

Part Six

Strong Relationships Make Dreams Possible

Part Six: Strong Relationships Make Dreams Possible

Green-Shifting Education
Raissa Marks

The laughter of my twin grandchildren echoes as I watch the rolling hills of Albert County from a comfortable chair. It's a crisp, beautiful, late summer's day, and this time next year the twins will be about to start kindergarten. They will enter the community-based education system that my colleagues and I envisioned long ago. I am proud of the battles we fought and the impacts of the work we did.

You may be too young to remember, but when I grew up, New Brunswick was stuck in thinking about a resource-based economy. Our leaders could never imagine a New Brunswick different from what it always had been. The model was simple: extract resources, sell for profit, and repeat.

The problem? None of it was sustainable. Even back then, it was clear the economic system had taken a toll. Hiking and walking, even as a child, I could see the hurt in the trees and forests, rivers and lakes, and plants and animals. I felt the toll on our communities. Our province had little money. Our health care system was strapped. We were mired in the debate about jobs versus the environment, which could never go anywhere.

Now, we have vibrant, innovative rural and urban communities. Our economy works in harmony with nature, not against it. Private and public investments support our young entrepreneurs. Social enterprises and co-operatives are economic drivers. When I think about what we have now, compared with what we had then, I see the shift began because of changes to our education system. We knew at the time that we were planting seeds for the future. Now, we can see the fruits of that labour.

When we launched the Sustainability Education Alliance in 2006, we worked with many organizations and agencies to develop a new vision for education in the province. We designed an education system focused holistically on interconnected real-world problems. We fought for a focus on teaching not facts but skills like critical thinking, problem solving, and collaboration. We wanted to foster community engagement in teaching and learning so that each student could reach their potential. It was no easy road to get where we are now.

I remember meeting an official from the Department of Education in May 2008 to discuss our vision for education. He was deaf to our ideas. He lumped us in with other interest groups and offered this putdown.

"We can't just change the curriculum for every Tom, Dick, and Harry from the Broccoli and Cauliflower Society that comes our way."

We had an uphill battle ahead of us. But we worked hard and things slowly started to change. Now the Sustainability Education Alliance takes a whole floor in the Green Shift NB building in downtown Moncton, alongside other organizations and agencies working to push green ideas forward.

One breakthrough came in 2014. The Conservation Council, Nature NB, Parks NB, the Groupe de développement durable du Pays de Cocagne, the Aster Group, and the NB Environmental Network imagined, planned, and launched a train-the-trainer service for environmental and outdoor education.

A few years later, the project became the *Great Minds Think Outside* program. The program secured funds for a few years, and it took off as teachers, principals, and parents began to see the value of outdoor learning. Students had improved mental and physical health; they worked better together; they could solve more complex problems; they had fewer behavioural issues in the classroom; they found a deeper connection with nature. The program found endowment funds from committed donors and has become a part of programming in every school, province wide.

Another breakthrough happened a few years later when the Minister of Education released his Green Paper on New Brunswick's education system. That paper caused quite a stir with some controversial outside-the-box thinking. Most importantly, it opened the door for new ideas to be brought forward and considered by the Department of Education.

People took up the opportunity. Response to the Green Paper on the education system went way beyond the ideas it presented. People began to talk about how sustainability could be fostered in the school system through community-based learning, outdoor learning, incorporating Indigenous knowledge, and teaching students to think critically about problems and solutions.

Part Six: Strong Relationships Make Dreams Possible

Eventually, things changed. Suddenly — or so it seemed — education became what we had envisioned so long ago. My grandchildren, the twins, will enter a system where students graduate from a world-class education system, where students leave to make their mark on the province and the world. Bright and innovative, students put nature and their communities first. They help create an inclusive New Brunswick, meeting everyone's needs, generating wealth, and having the highest Gross Happiness Index in all of Canada. What a difference the green shift in education has made in New Brunswick! What an accomplishment!

Letters from the Future

Taking Back our Resources
Chris Rouse

I'm old but happy sitting on my deck looking out from the Kingston Peninsula at the beautiful and bountiful Wolastoq River with the love of my life, whom I met amidst the tumult of the early 2020s. I will never forget that time, not only because of the momentous world events, or because I met my wife, but also because a chain of events began that led me to become the CEO of Wabanaki Energy, formerly known as NB Power. Through Wabanaki Energy, New Brunswickers and First Nations took back control of our shared resources from private interests for the benefit of everyone and not just the few.

I met my wife when the Council of Canadians hosted a community workshop, which I facilitated in Fredericton. We developed an Integrated Resource Plan for New Brunswick at that workshop, which enabled our province to meet and exceed the Paris Climate goals. As we sit here on our deck, we are happy to think about how this plan helped New Brunswickers come together to meet these goals. What we agreed to that day evolved into what was later adopted in all the Canadian provinces and territories, and by other countries around the world.

We held the workshop after the Green Party's changes to the *Electricity Act* were defeated in the legislature. After the vote, I approached the Green Party, First Nations, and environmental groups about collaborating on better legislation that would put New Brunswickers back in control of our publicly owned utility, NB Power. The workshop helped us find common ground and gave us something to fight for. We won the fight against fracking, pipelines and nuclear generation, and now dedicated our time to building the energy system of the future our province so desperately needed.

The first thing we did at the workshop was to discuss how to give New Brunswickers and First Nations more control of their energy in the future. We also discussed the privatization of our renewable energy resources. After much discussion, people decided against giving away our resources to capitalists to profit from. Instead, a model of public ownership was adopted so all New Brunswickers and First Nations would become investors in renewable energy. Delegates then turned to evidence-based, least-cost, environmental, social, and economic sustainability principles to guide the decision-making process. Each decision was consensus-based and resulted in a new Integrated Resource Plan for New Brunswick. The plan, which enabled First Nations to partner with NB Power in the development of renewable energy resources, would lead to an equitable solution for everyone in the province.

The Green Party formed a minority government after a dispute over language issues between the People's Alliance and the Progressive Conservative Party led to a non-confidence vote and a new election. The Green Party introduced new legislation based on the new Integrated Resource Plan and it passed.

The revenue from the Carbon Tax was invested in renewable energy. The return on this investment was, in turn, reinvested in the continual development of renewable energy. This investment had a powerful economic compounding effect, providing capital needed for the transition to a clean energy future. The response of the economy to this on going investment strategy served to fund social programs, education, community services, lower taxes, and even helped pay down government debt.

My wife became a Green Party MLA and was appointed Minister of the Environment. She refused to sign any more permits for spraying New Brunswick forestlands. This led a total revamping of provincial strategy based on the restoration of Acadian forests.

I was appointed CEO of NB Power. The first thing I did was cut my salary by 50%. I expanded efficiency subsidies for low-income homeowners and cancelled the rest. We used the savings to create an efficiency investment program in solar panels, home retrofits, and electric cars. Unlike earlier programs, these new programs were targeted so all New Brunswickers could participate instead of only those who could afford to participate.

We created community-based programs to support locally created and community-operated renewable energy projects. These projects created good paying jobs and supported local businesses that designed, built, installed, and maintained the new, publicly owned, renewable energy infrastructure.

One of my most memorable moments as CEO of New Brunswick's publicly owned utility — now renamed, Wabanaki Energy — was the day the Point Lepreau Nuclear Power Generating Station was shut down. I had the honour of pressing the button to initiate the final shutdown of one of the worst decisions in NB Power history. Ironically, decommissioning the plant created many new jobs. Several New Brunswick businesses became a hub for developing and then exporting decommissioning engineering services. Indeed, this decommissioning industry ended up being the only profitable aspect of the nuclear energy industry in New Brunswick.

Just before I retired, my last achievement as CEO of Wabanaki Energy was decommissioning the Mactaquac Dam. The beautiful and bountiful Wolastoq River restored itself and began to heal. Wabanaki Energy had achieved zero emissions ten years ahead of schedule. Renewable generation and storage technology has become so efficient and cost effective that we didn't need the dam.

As I sit on my porch and look at how beautiful the river is, I will never forget that day when the fifteen First Nations and I signed the contract to decommission the dam, just as we had agreed at that summer workshop. It gave renewed meaning to our Peace and Friendship treaties.

Part Six: Strong Relationships Make Dreams Possible

The Joys of Rural Work
Teri McMackin

This late summer morning is unseasonably cool and damp. As I wait for the coffee to finish steeping, I slip on rubber boots and wander out to the garden to look at the peas. The dew shimmers as the sun rises. Boots were a good choice.

The dog runs up to me, having just chased another deer away from the plot. He's proud of himself, though he has nothing to show for his accomplishment other than a sloppy goofy grin. He's getting on, but if you didn't know him, you wouldn't see it. With what I imagine is a puppy's joy, he greets our four old hens—the most we're allowed within the Petitcodiac village limits. The hens have outlived their egg-laying days and have been promoted to pest control. The kids cannot fathom eating them; "They have names, you know."

Back to the house, I pour the coffee black and read the local *News & Views* paper. An advertisement promotes the village's fourteenth Arts and Cultural Festival on the front page. It reminds me, I must bug Council to send out a notice to local businesses that Main Street will be closed early that day to give vendors time to set up.

The festivities have come a long way since our modest Kay Street Party ten years ago. People come from all over, the bed-and-breakfasts fill up, and the campground, which opened five years ago near the community garden, catches the overflow. The festival is a popular space, even more so since the local solar co-op opened in the neighbouring field a couple years back—one of the first of its kind. Almost everyone taps into the solar panels, which have become a model for local cooperative energy production and distribution.

I drink my coffee and read a few articles about upcoming community events before going to hang the laundry on the clothesline before everyone wakes up. The village had an initiative a couple years ago to get more people to use their clotheslines. The village bought them in bulk, and the works department offered free installation for anyone who wanted one. Uptake was great. Almost eighty percent of people in Petitcodiac regularly use their clothesline instead of their dryers, especially in the summer months. Every little bit to reduce electricity usage, they say.

Ethan, my son, is the first one up. He has just finished grade ten. The school received an upgrade a few years ago: it added a shop for welding, mechanics, and renewable energy. The trades were always popular with the kids in the village, but with the new solar farm, the school has added a program focusing on training in renewable energy technology and systems. Ethan is excited to be in the first cohort of students to enter the program.

Ethan is tall and lanky, like his father. He sits down at the table as I place a plate of blueberry pancakes in front of him. If it weren't for the garden, he and his brother Logan would have to work part time just to buy the food they eat. Thankfully, our freezer is still stocked with blueberries from last year. Logan stumbles into the kitchen, half asleep. My husband is close behind. Logan will be entering high school in the fall. The three are hungry and grateful that I am a morning person and that the pancakes are already coming off the griddle. Once everyone eats up and then cleans up, the two boys are out the door.

Believe it or not, my boys still kiss me goodbye. They are off, with an "I love you, Mom." Their friends are waiting on bikes at the end of the driveway. The boys have a tight-knit group of friends. I was a proud mother when Logan came home from school last week with a new friend. Logan loves reaching out to meet people. It's never easy being the "new kid." His family had moved here from Ontario because living in our small village is simpler and far more affordable. Our village saw an influx of new families when the co-operative housing development went in just off the Old Post Road.

With the kids off on their bikes, my husband and I head to our office above the kitchen. We both work online. The Internet was already pretty good when we first moved here sixteen years ago, but not everyone was so lucky. Many people outside the village had limited access. But since then, community upgrades have ensured high quality, dependable connections. Everyone who wants to can get online, and fast.

My husband turns on his computer and puts on his headset. I tell him not to bother with the headset as I pack my laptop bag. I'm going to work

Part Six: Strong Relationships Make Dreams Possible

from the Main Street Café for the morning. I grab my bag, kiss my husband, jump on my bike, and head down to Main Street.

The Café is in one of the oldest buildings in the village, one of a few original buildings that have survived. It needed a lot of work, but the young couple that moved here from away took great pride in restoring it. The Café is a charming and friendly place, a hot spot for local online workers like me.

Over the last five years, a dozen new businesses have opened in the village. They employ almost a hundred people. Some have taken up space on Main Street, which is now so full that one local developer purchased some land near our new Medical Centre to construct a LEED-certified building for another two businesses. People used to have to travel to Moncton or Sussex to look for work, but now there is not so much need to leave our small village. All the businesses sell locally, but they also use the Internet to reach wider markets. We have everything we need right here. Many "meetings" are held in our little Café, I like to joke. Petitcodiac's Main Street Café is where so much magic happens. It's our commons.

I put the bike in one of the designated bike spaces, enter the Café, sit in my favourite corner where there is an electric outlet, and ordered tea—rather than more coffee—and a scone.

The woman who makes the scones sells them at the Market on Saturdays and also to the Café. I don't need to wait until Market Day for a good scone fix. The tea and scone soon arrived at my table. The young man working as a *barista* had already begun preparing my order when I walked through the door—the perks of being a regular.

As I pick away at my work, people come and go. I welcome the odd interruption of a passing conversation. Sipping tea and slowly eating the scone, I type, I think, I type. The "white noise" of the busy Café is the soundtrack for a joyful and productive summer morning in rural New Brunswick.

Letters from the Future

Local Food, for Local Communities
Stephanie Coburn

As I write to you from our farm near Head of Millstream, it is indeed a joy to see the work of many years come to fruition. I can step outside my door and see five farmsteads on a piece of land once the possession of just one family—five homestead farms producing food for fifty families, all within 20 kilometres.

Such strides have been made in renewable energy systems that the caloric value of our food system now is ten times the calories used in its production. And the regulatory systems have also advanced to favour local processing for a local market.

We have a fruit farm producing apples, pears, peaches, raspberries, strawberries and other small fruits. In addition to providing fresh fruit in season, the family run business turns their fruit crops into juice, jams, cordials, and wine.

We have a dairy farm, powered by solar energy. The cows and goats on that farm give us milk, cream, butter, cheese, and yogurt. The residues from the processing are sent to another homestead where they are combined with the fruit residues to feed pigs enjoying the fresh air and sunshine in their outdoor enclosures. Nearby are fields planted with oats and peas to be made into silage for the pigs, cows, and chickens in the winter.

Companions to the pigs are the chickens, also enjoying the fresh air, bugs, and green grass as their pens are moved across the field, leaving their residues to fertilize next year's hay crop. The hay feeds the cows, goats, and horses during the winter. The horses work in the summer to till and cultivate the vegetable gardens and in the winter to yard out the wood from the woodlot to heat homes and greenhouses. One of the homesteads produces the vegetables for our fifty families. They have garden space for the outdoor crops in the summer, and greenhouses to provide fresh produce all winter. The greenhouses are heated geo-thermally with wood heat backup on particularly cold winter nights.

We send boxes of food to our families all year long—fresh vegetables spring, summer, and fall and storage vegetables in winter: lettuce, spinach, beet greens, chard, kale, zucchini, beans, corn, peas, tomatoes, peppers, eggplants, broccoli, cabbage, pumpkins, potatoes, onions, carrots, squash, parsnips, and turnips. The only things we can't grow are okra and avocados!

Our beef farmer also has movable chicken houses for the laying hens who dash outside as soon as the doors are open in the morning to eat grass and chase bugs. They lay the most delicious eggs we've ever tasted. The families in this community all work together in the fall to store the bounty of the summer for the cold winter months.

All over New Brunswick we find this kind of homestead farm producing food for their neighbours. Instead of sending 96% of every food dollar out of the province for food grown somewhere else, we keep those dollars circulating in our communities. Government has stopped subsidizing huge farms and is helping young people set farm operations that contribute to our self-sufficient and sustainable local food systems.

In our larger community of fifty families, we have a doctor, a dentist, several nurses, a mechanic, an electrician, a plumber and several carpenters. About ten years ago we started growing hops—a very hardy crop—and, lo and behold, a brewer set up in town and now we have our own local beer as well.

As I sit on my doorstep and think about the last few decades, I'm so happy we elected people to all levels of government with the vision to think about what was possible for New Brunswick and who worked to open the possibilities for local food production and community building. "Without vision, the people perish." But with the vision of a self-sustaining community and the political tools to help, changes were made that produced this result. We knew it could be done, and we did it!

Letters from the Future

Gardening the Margins
Kylie Bergfalk

It's 8:35 a.m. on an unusually cool morning in early August. A small group of people have gathered at the corner of George and Westmorland Streets in Fredericton. We're a mixed group: teenagers tote hoses and spades, a couple of children arrange races between fire hydrants, retirees in sensible shoes exchange notes on summer travel plans to New Brunswick's provincial parks, and working-age folks trickle in — the sort who ten years ago would have been facing computer screens in offices on a Friday morning. But since the 20 hour work-week has become common, the meaning of "work" has shifted and broadened. Community spaces and services like our Friday morning gathering have become sites of intergenerational activity and a new kind of "work."

On mornings like this one, groups like ours gather in neighbourhood corners all over the city. People come together to observe, to care for, and to harvest from Fredericton's "garden margins," which dot the city. People are cared for in return. Bees and butterflies have already been busy for hours this morning. There will be songs, sweet peas, and late-season strawberries warmed in the sun in the hours ahead.

Fredericton's community gardens have changed radically in the past decade. Instead of just small plots tucked away in abandoned spaces between streets, suburban yards, and on school and university campuses, the entire community is now a garden. People have learned to participate in the urban ecosystem in ways that acknowledge and enrich the relationship between the human and the more-than-human world. The restoration is reciprocal, co-created, and jointly enjoyed. It turns out a lot of edible and pollinator-friendly plants can be grown in the urban margins where once were only parking lots and on the edges of Fredericton's sun-exposed curbs.

The purples of the drought-tolerant hyssop and wild bergamot appeal to the eyes of both humans and hummingbirds. The blue vervain along the riverbanks and throughout much of the flood-prone downtown will be a source of winter food for the songbirds. Herbs freed from pots and

Part Six: Strong Relationships Make Dreams Possible

kitchen gardens—basil, coriander, thyme, mint, and sage—proliferate on every block to the delight of cooks and bumblebees alike. In places where the soil quality remains poor, black-eyed Susans cheerfully flower through the summer months. Clover and yellow cinquefoils provide groundcover and the promise of richer soil in years to come.

In Fredericton, in early August, on a Friday morning the streets are a riot of colour—the garden margins are everywhere in bloom. It's not quite an idyll, though. In the face of environmental disaster, the shift away from carbon-intensive agricultural and the new dietary habits have been a bumpy transition brought on by the urgent need to decarbonize. It was as much forced, as it was chosen. There are no lawns anywhere.

Water, oil, and food rationing made it seem insane for citizens and cities to maintain green grassy spaces in the usual way. At first, the loss of lawns was seen as a misfortune, one more blow of almost too many to bear in an increasingly unpredictable world riven by crisis and steeped in despair.

But in Fredericton, in the midst of crisis, hope has found a place. It was the historic, all-women City Council, in consultation with the Wolastoqiyik Elders, that implemented the citywide program of gardening the margins. Gardening the margins has linked ecological restoration with local food systems and with citizens' mental, emotional, and physical wellbeing.

Neighbourhood work groups were organized to take on the labour of learning and restoring—and the labour of learning and restoring we did. As Robin Wall Kimmerer wrote in *Braiding Sweetgrass*:

> *Restoration is a powerful antidote to despair. Restoration offers concrete means by which humans can once again enter into positive, creative relationship with the more-than-human world, meeting responsibilities that are simultaneously material and spiritual. It's not enough to grieve. It's not enough to just stop doing bad things.**

Now in Fredericton, on this morning and on many other mornings, we are following the example of natural ecosystems around the world. We are learning to manage precarious circumstances and uncertainty with diversity. When many species play their roles in an ecosystem, even when facing unpredictable conditions, there is more support for all.

When young and old and everyone in between, both human and more-than-human, gather in Fredericton's many neigbourhoods on mornings like this, restoration is happening from the ground up—another moment of hope among many.

* Kimmerer, Robin Wall 2011. *Braiding Sweetgrass: Indigenous Wisdom, Scientific Knowledge, and the Teachings of Plants.* Minneapolis; Milkweed Editions.

Letters from the Future

Blessings from 2066
Jean Desrosiers

I am Wewo. I live in the 5th commune of the town of Nigadoo, which signifies "good place to hide" in the Mi'kmaq language. At 99, I've lived one third of my life in the 20th century and two-thirds in the 21st.

I am lucky enough to have my health, but mostly I am blessed to still be sharing everything with Sadhana, my companion of 73 years. We live autonomously in a bright and beautifully furnished co-op apartment in the heart of the commune. The co-op is run entirely by volunteers. Pretty well every resident of the commune is a member of the co-op, and the profits of the business go toward improving the shared circumstances of our lives.

The most wonderful thing about my life is that I am never alone unless I want to be. I ponder with great sadness how lonely the elderly must have felt in the past, tucked away in retirement homes and cared for by underpaid, underappreciated staff. Today, we live in mixed age communities; young adults are always ready to lend a hand, children of all ages come and go, bringing us gifts and begging us for stories. And we do have lots of stories. Most of them are happy ones. Let me tell you one right here.

In the late 20th century, the government of New Brunswick started talking about full municipalization. The principle was that every settlement area in the province should either be incorporated as a self-standing entity or annexed to an existing city, town, or village. The idea was met with fierce resistance by folks who lived in the unincorporated rural areas and were terrified at the thought of seeing their taxes double or triple, while services would likely decline and potentially crumble. There was much turmoil within the rural communities though very little of it made headlines. Most of us had the feeling that municipalization was going to happen despite our fears and on our backs.

Then Ti'tigli—whose name was Alex back then—returned home from their trip around the world and became our advocate. Ti'tigli had a combination of discerning wisdom and tireless energy that we often see in the same person over the course of their life but rarely at the same time. They had a capacity to listen and to extract what all parties wanted from stances that seemed poles apart to most people.

After seemingly endless negotiations that spanned eleven years and three governments, a consensus was finally reached that would rest on three core principles, two of which would be enacted into legislation.

First, the Fair Tax Act ensured that people who lived further away from central services such as libraries, sports complexes, and health centres would pay a lower tax rate to compensate for the lesser availability of these services and the travel that would be required to access them. This rested on a complex but comprehensive formula that ensured levelized taxation.

Second, the Land Occupancy Act was also passed to financially encourage landowners with wide road frontage to subdivide the front of their property for the construction of more residential buildings. Why own three hundred metres of road frontage when they only needed twenty metres to access their land? The newly subdivided lots were offered for sale to families of all financial means, at prices based on their capacity to pay.

Sparsely settled areas, such as the one that includes the formerly named hamlets of Sormany, Nicholas-Denys, and Free Grant, saw rapid population growth. This in turn made providing many essential services feasible, including high-speed internet, a general store, a nurse practitioner's office, and an emergency services complex. At first, some folks said these services weren't needed when a large town was nearby. But with an aging population and a drastic reduction in the use of motor vehicles, the proximity of these services was paramount to the vitality of our community.

The co-operative also built a beautiful community services facility on the south bank of the Millstream River, along which many well-to-do businesspeople once had snowmobile camps when I was young. An array of holistic health professionals, such as acupuncturists, massage therapists, and osteopaths, are now housed in this building. Since these professionals use the office space on a shared basis, new services can still be added.

One of the key aspects of our commune is that we don't have many "assets" sitting idle. The same goes for all our rooftops; I challenge you to find one without solar panels. But I'll let someone else tell that story.

The third and most important of the principles, however, was the most difficult to implement, to the point of appearing almost insurmountable. This principle could not be enacted into law; it had to be adopted and embraced by the population. At the time, Ti'tigli called it the "unity in spirituality principle." The crux of it was that all persons, no matter their religion, skin colour, language, customs, dress, sexual orientation, gender — no matter what — were equal. We are all the same, creatures that require love and acceptance to thrive. The only traits we would not accept were greed, intolerance and prejudice. Similar initiatives were sprouting all around the world under many different names. The timing was perfect because we soon started welcoming a multitude of climate refugees from all parts of the world.

Our ability to make newcomers feel safe and at home was the ultimate test of this principle, and the result was an unequivocal success. Young people now never even notice that some people have different colours of skin. This culture of acceptance and kinship gave way to beautiful friendships and love stories, which led to much blending. We all express our faith and spirituality in different ways. Some folks believe solely in science, but the one unifying characteristic is that we all love and respect each other, no matter how different we may appear at first glance.

In time, the population of Nigadoo has quadrupled, keeping pace with the rest of the province. This is quite a reversal from the loss of population Northern New Brunswick was experiencing in the first decades of the century.

As a community, we now grow more than 75% of the food we eat. We have geo-thermally heated vertical greenhouses and many forest gardens. We eat so well! We usually make a simple raw breakfast in our quarters, have lunch outside every day the weather permits and then have dinner in the common room.

We no longer do animal farming, as we have all become vegan. It happened gradually over decades through a combination of conversion and attrition. As the years went by we came to the realization that enslaving and killing animals for sustenance is not necessary. If you wonder how this is possible, I invite you to seek out the literature on the subject and look within your heart for the answer. May we come to see the utter joy of reverence for life and act with care for all that is living.

Afterword:

Change the Story, Redefine Progress, Create the Future

In October 2018, the Intergovernmental Panel on Climate Change (IPCC)[1] gave the carbon polluting nations of the world twelve years to make dramatic changes to our fossil fuel dependant way of life before the impacts of climate change become catastrophically irreversible. Three years after that warning, the momentum for the changes needed is painfully slow and in some cases simply absent.

With the Paris Agreement in 2015, the nations of the world participating in the UN Convention on Climate Change signed on to targets for voluntarily reducing their carbon emissions, an encouraging sign the climate crisis was no longer being ignored. But the IPCC's 2018 twelve-year timeline for accomplishing these reductions was "a shot across the bow" of every signatory nation to the Paris Agreement.

The world has gone from climate change to climate crisis to climate *emergency*. Time for a mitigating response is running out. The speed and scale of the action required is now compared to the response of Britain, Canada, and the US to the Second World War; virtually overnight they repurposed their economies. That's the scale and scope of the action the climate emergency requires. Fossil fuel powered, unlimited economic growth is an addiction we must break in order to preserve a habitable world. The business-as-usual old guard, reaping profits from unsustainable growth, and their political allies, are slow to *get it*. But the school children and young people of the world are not; more and more, they *get it*!

In the spring and fall of 2019, students in the villages, towns, and cities of New Brunswick joined their peers worldwide in climate action strikes—

Fridays for the Future. Some adults marched with the youth but many were unsettled to see students take to the streets on Fridays instead of the classroom. They were also troubled to see the placards and hear the chants accusing the political and business establishment of ruining the future for the world's children. But more shocking was hearing teenagers anxious about an earth so damaged by climate change that widespread extinction is likely and the end of life as we know it a real possibility.

When we were teenagers, the prospect that climate change could devastate life on the planet was not something that disturbed our minds. Now, young people are distinctly aware they are inheriting a climate catastrophe: previously unknown temperature spikes with persistent, lethal heat waves; prolonged drought with crop failures and food system collapse; melting glaciers and icecaps leading to coastal sea level rise where the majority of earth's human population is concentrated; massive forest fires that often outstrip fire fighters' control; increasing frequency, intensity, and geographic scope of powerful storms sweeping across landscapes, smashing homes and businesses, and tragically disrupting the lives and livelihoods of families and communities; destructive flash floods from unprecedented rainfall in short periods of time; waves of climate refugees driven by poverty and violence who continue to seek sustenance, safety, and, shelter wherever they can find it.

All this is now happening. Lacking an effective reduction in greenhouse gas emissions, all this will get worse. Life as we knew before the climate crisis is gone and a terrifying change is underway—it can be a paralyzing prospect. Young people, especially, understand what is happening and are shouting for the old guard to stop doing business-as-usual and act to address this worldwide emergency.

* * * *

Times of crisis can trigger transformation. There are encouraging signs in New Brunswick. Voters wanting the province to take strong climate action elected three Green Party members to the New Brunswick legislature in 2018 and re-elected them in 2020. In 2019, voters in the Fredericton riding elected their first woman and first Green Party member to the Parliament of Canada. In February 2019, Edmundston was the first city in New Brunswick to declare a climate emergency, followed by Moncton and Saint John and then other communities.

Afterword

All these developments signal a hunger for change. Topics once impossible to imagine being discussed are now part of everyday conversations and mainstream policy debates. But to transform our economic and social worlds in a way that would stop a climate catastrophe, we need more than protests and student strikes, as important and heartening as they are. To build a better future we need new ideas leading to actions that create sustainable economic and social relationships. Fighting against what has gone wrong only goes so far. Fighting together to rescue our communities and the world from the ravages of climate change requires achieving a new kind of progress — the kind of progress recounted in the letters from the future in this book. Positive, shared narratives are essential.

And the sources of these narratives need to be diverse. The stories that shape our future must not come from just the usual suspects who have been controlling the provincial narrative for decades through their virtual monopoly of media outlets and their red and blue political dynasties. We need to hear the voices marginalized from the corridors of power and corporate news: Indigenous people, women, racialized people, immigrants, artists, queer and trans folk, people living on the edge, and youth; they will be faced with enormous challenges in the coming years. We must listen to what they are telling us.

Cultural analyst, Evan Watkins, asks how those holding the controlling positions of power in market societies maintain their control? His illuminating answer — *by controlling the narratives of change*.[2] We live in an environment of dynamic change. We are constantly told by market based advertising and political storylines about how things are changing or should be changing. We are constantly fed narratives on what we need to buy, buy into, and do to keep up with the changes the market economy presents as progress. This progress, however, is largely illusory but its narrative serves those who control market societies and are determined to maintain their control.

Two examples from our experience in New Brunswick — one past and one current — illustrate how controlling the narratives of change works. A few years ago we were told in no uncertain terms that New Brunswickers must accept maximum use of fracking for the production of natural gas in order to provide the province with a prosperous, modern economy. Industry advertising and government exhortation presented this narrative of change in glowing terms. The government fully expected the promise of jobs in rural New Brunswick and the prospect of increasing natural resource royalties for the province would seal the deal.

As it turned out, hydrocarbon prosperity for New Brunswick was not the only narrative in play. A coalition of rural residents from the areas targeted for exploitation, environmental conservationists, Indigenous people, technology researchers, social scientists, and concerned citizens fought to promote an alternative narrative. The coalition exposed the environmental damage and negative health effects associated with fracking. Its analysis discredited the government's storyline that New Brunswick's potential for hydrocarbon production would dramatically boost provincial prosperity. Anti-fracking demonstrations and blockades followed. The industry backed away from further exploration. A new government was elected and put a moratorium on fracking. The government changed hands again and the moratorium stayed. Fracking in New Brunswick now appears to be a dead issue.

One outcome of the coalition against fracking is the Peace and Friendship Alliance of Indigenous peoples and settlers that continues to meet today to build our shared understanding of what it means to all be Treaty people.

The narrative advanced by the coalition against fracking wove another important element into its presentation — the rapid advance and falling cost of renewable energy technology. This was a key storyline in the coalition's narrative of change. The sun is setting on the hydrocarbon industry. Even Mark Carney, the former Governor of the Bank of Canada and Governor of the Bank of England at the time, warned investors about the stranded assets, high debt load, increased risks, and diminishing prospects of the fossil fuel industry. The narrative of change for expanding the hydrocarbon industry failed in New Brunswick, even as a narrative of change based on renewable energy advances.

The current example of competing narratives of change in New Brunswick comes to us from the nuclear energy industry and its opposition fighting for renewable energy. Two separate start-up operations have come to New Brunswick with proposals to build small modular nuclear reactors (SMR) at Point Lepreau. They have been successful in their requests for subsidies from both the New Brunswick and Canadian governments, which they claim are essential for attracting private investment. The coalition that defeated fracking has now become the Coalition for Responsible Energy Development in New Brunswick (CRED-NB) and is challenging the story being told by these start-ups and their allies in business and government.

The New Brunswick supporters of nuclear energy have revived the narrative of change that was first used when the large nuclear power plant was built at Point Lepreau between 1975 and 1983. They imagined the Saint John region would be become a major energy hub for the Maritimes and New England. It didn't work out that way. Supporters of nuclear power now believe this development can be realized with a cluster of small modular nuclear reactors at Point Lepreau.

The Coalition for Responsible Energy Development is once again calling up its story of renewable energy technology as the progressive path of change even as it points out the unacceptable risk and long-term environmental toxicity of nuclear power. In addition, the decade long timeline required for the development of this technology and the high level of uncertainty that it will prove viable, and if viable, commercially competitive, make these attempts to revive nuclear power a foolish investment.

The rapid advances in renewable energy technology, its decreasing costs, its immediate availability, and its relatively quick investment payback time give the Coalition's narrative a commanding advantage over the nuclear power story. It remains to be seen whether the Governments of New Brunswick and Canada will continue to invest tens of millions of dollars in an unproven, high risk, and environmentally toxic way of generating electricity, when a proven, widely deployable, increasingly affordable, and safe alternative is immediately available. The narrative of nuclear power has a high tech mystique that appeals to a certain remnant of political figures and business interests. But it is the story of renewable energy that is gaining the attention of homeowners, business people, municipalities, and political leaders who understand the need to define a new kind of progress for New Brunswick.

* * * *

The old narratives no longer work. The domination of profit-maximizing multinational corporations, the climate crisis, and now a pandemic have damaged communities and threatened livelihoods. Cheap commodities and entertainment attempt to lull us into being passive consumers, feeding the wealth accumulation of a neo-colonial globalized capitalism that plunders the planet for profit. Our high levels of consumption are made possible by cheap and polluting hydrocarbons, the root cause of the climate crisis. Colonialism has abandoned its Imperial robes and put on a business suit; it has shifted from political rule to economic domination. Now called Neoliberalism, its environmentally destructive and socially disregarding behaviour is, none-the-less, intellectually, ethically, and morally bankrupt.

New narratives are needed now. *Letters from the Future* seeks to share these narratives at this critical time in the history of New Brunswick and the whole planet. Effective narratives of change locate the problems of the present in the inherited order of things and offer solutions that create a more equitable and ecologically sound world.

In response, we see the stirrings of widespread rejection of globalized, neoliberal, market-driven domination. A new narrative for the future is being propagated. *Letters from the Future* is evidence that New Brunswickers are engaged in developing this new narrative. We hope our readers will be moved to reflect further about how progress can be redefined as enhancing life instead of maximizing profits and accumulating wealth. We hope readers will take inspiration from the letters that describe ways of life dedicated to building equitable, inclusive communities and prosperous, resilient local economies.

We take seriously the credo, "think globally, act locally." We need to learn how to live sustainably from the resources in our communities, rather than importing destructive mining operations and nuclear power plants that enrich only their owners and shareholders who live elsewhere, while leaving New Brunswickers with ravaged landscapes and toxic legacies for future generations. If we can do this, our accomplishments as a society will stand as an example to others communities and other jurisdictions around the world, just as others who are accomplishing similar transformations inspire us to press on for the better world we know is possible.

* * * *

Now, as we write this Afterword in mid-2021, we are increasingly stunned by reports of new climate change extremes: heat parched crop failures across the Canadian prairies; another season of destructive forest fires up and down the west coast of North America, northern Ontario, and across Siberia; an anomalous heat dome spiking temperatures in British Columbia to levels never thought possible; a two month rainfall in twenty-four hours over parts of western Europe pushing flash floods with no precedent on a rampage of destruction through heavily settled areas; the water crisis in drought stricken areas with reservoirs on which whole urban regions depend dropping to a small fraction of their capacities with no replenishment in sight. Add to this the lingering and now resurgent threat of the COVID pandemic due to new strains of the virus, a fragile economy, plus increasingly volatile conflicts between powerful nations and within nations, even in the United States.

Pessimism is easy, even understandable, in face of these global circumstances. What can we in New Brunswick do about these earth-spanning issues? *Letters from the Future* emerges out of a shared conviction that, for a start, we need to replace the market-based, globally focused economic growth narrative that dominates our time. We need a flowering of new, innovative narratives that create and nurture local initiatives for real, long-term solutions for sustainable living.

Letters from the Future is a book of *speculative nonfiction*; the authors are imagining the changes they would like to see and the accomplishments that flow from those changes, but they are writing about the real world, the world we live in, not a fictional world. Near the end of his life, one of the 20th century's wisest economist and social scientist, Kenneth Boulding, wrote the following about the importance of speculative nonfiction.

> *The great predicament of the human race is that all experiences are of the past but all our decisions are about the future. Unless we at least think we know something about the future, decisions are impossible, for all decisions involve choices among images of alternative futures. This is why the study of the future is more than an intellectual curiosity; it is something that is essential to the survival of humankind itself.*[3]

This is the study that *Letters from the Future* has undertaken; it is the continuing study, leading to action, that we are invited to take up as we redefine progress, confront the climate crisis, and do everything we can to create the better world we know is possible.

The Editors

References

1. Intergovernmental Panel on Climate Change, "Special Report: Global Warming of 1.5 °C," 2018, https://www.ipcc.ch/site/assets/uploads/sites/2/2019/06/SR15_Full_Report_Low_Res.pdf

2. Watkins, Evan, 1993. *Throwaways: Work Culture and Consumer Education*. Stanford, CA: Stanford University Press.

3. Boulding, Elise and Kenneth E. Boulding, 1995. *The Future: Images and Processes*. Thousands Oaks, CA: Sage Publications.

About the Authors, Editors, & Illustrator

Tom Beckley is an environmental sociologist at the University of New Brunswick and a forest steward at Keswick Ridge.

Kylie Bergfalk is a writer who lives in Fredericton.

Stephanie Coburn is a farmer, gardener, mother, and grandmother. Her farm is near Head of Millstream.

Victoria Clowater (they/she) is a graduate student at McMaster University studying how public policy affects queer and polyamorous Canadians. Their New Brunswick roots are in Fredericton, Saint John, and the Miramichi River Valley.

Alain Deneault is a Quebecois philosopher who lives on the Acadian Peninsula of New Brunswick. He is professor of philosophy at the Université de Moncton on the Shippagan campus. He is a member of the Collège international de philosophie in Paris.

Jean Desrosiers is an Industrial Technology Advisor for the Industrial Research Assistance Program of the National Research Council of Canada. He lives and works in Bathurst.

Carl Duivenvoorden is a speaker, writer, sustainability consultant and dad living in Upper Kingsclear.

Jael Duarte is lawyer specialized in immigration, refugee, and family law. She is a newcomer to New Brunswick from Bogotá, Colombia.

Sagda Elnihum immigrated to Canada from Libya six years ago. She is a graduate student at the University of New Brunswick where her research focuses on the discrimination that immigrant women face in Canada.

Amy Floyd lives in Taymouth, New Brunswick. She works in food security, permaculture, and rural community development.

Naomi Gullison is a trans-woman and advocate for increased post-secondary funding. She was born in St. Stephen and moved to Fredericton to attend St. Thomas University and pursue a degree in Native Studies.

Mathew Hayes is professor of sociology and Canada Research Chair in Global and Transnational Studies at St. Thomas University.

LA Henry is a criminal and family law lawyer practicing in Fredericton. She is the Executive Director of the Fredericton Legal Advice Clinic and an instructor at UNB's Faculty of Law and at St. Stephen's University.

Cheryl Johnson lives in rural Upham, teaches math, and lives off the land as much as possible.

Lauren R. Korn holds an M.A. in poetry from the University of New Brunswick. She currently lives on the aboriginal territories of the Salish and Kalispel people (Missoula, Montana), where she is the Director of the Montana Book Festival and the host of Montana Public Radio's literature-based radio program and podcast, *The Write Question*.

Nigam Khanal moved from Nepal to New Brunswick for her graduate studies in 2012. She currently lives in Fredericton and is a founder of the Immigrant Women's Association of New Brunswick (IWANB).

Renelle LeBlanc is the Executive Director of Aulnes Projects, a not-for-profit media production co-op focused on community development in rural New Brunswick. She lives at Nigadoo.

Louise Livingstone is an ecologist, journalist, and former farmer who used to live on a hundred-acre farm in Eastern Ontario, who now lives in Gagetown.

Abram Lutes (editor), originally from Woodstock, is a graduate student at Carleton University in Ottawa.

Raissa Marks was born and raised in the hills of Albert County, New Brunswick and currently lives in Montreal with her family.

About the Authors, Editors, & Illustrator

Mary Louise McCarthy-Brandt is a diversity educator and poet who is dedicated to telling the stories of her ancestors, the early African people of New Brunswick. She lives in Fredericton.

Teri McMackin is a career volunteer, actively involved in community development initiatives and local government. She is a self-employed digital independent and ghost writer.

Arianne Melara Orellana is a Youth Leadership and Education Policy advocate representing the voice of young people from diverse backgrounds. She is an immigrant from El Salvador and Board member of the Atlantic Summer Institute and the NB Champions for Child Rights.

Carlos Morales is a poet and writer from El Salvador who has lived in Fredericton since 1991, where he participates in literary activities and works to promote Latin American culture.

Alicia F. Noreiga is a Black international doctoral student at the University of New Brunswick, where she studies rural education. She was a primary school teacher in Trinidad and Tobago for nineteen years.

Susan O'Donnell (editor) is a researcher, writer and activist and lead investigator of the RAVEN project at the University of New Brunswick.

Ajay Parasram is an Associate Professor in the Department of International Development Studies and the Department of History at Dalhousie University in Halifax.

Adje Prado lives in St-Joseph-de-Madawaska and is a specialist in environment and climate change adaptation.

Chris Rouse is a New Brunswick industrial systems integrator and the Founder of New Clear Free Solutions, an environmental group that provides objective scientific, financial, and regulatory information to the public and government decision makers.

Terry-Ann Sappier is a Wolustukeyik from Negoot-gook, a mother of four, a grandmother of six, a land defender, and an environmentalist.

Erin Seatter is an award-winning journalist who lives and writes in Vancouver.

The Skutik communiqué was written collaboratively by the Passamaquoddy Recognition Group, Maritime Social Innovation Lab, along with Art MacKay and Kim Reeder.

Ian Smith (illustrator) is a grandfather, Outward Bound Canada Instructor, retired Parks NB Program Manager, outdoor educator, and artist. He lives near Harvey Station.

Sara Taher is an Egyptian blogger and a small business owner living in Fredericton. She works as a technology specialist and volunteers for the fundraising committee of the Fredericton Community Kitchen.

Lisa-Gay Taylor, a recent immigrant, is a research student and advocate for immigrant women She lives in Fredericton.

Daniel Tubb (editor) is an environmental anthropologist at the University of New Brunswick. He lives in Gagetown.

Leland Wong-Daugherty makes kites in South Knowlesville, where he and Tegan Wong-Daugherty have established the Knowlesville Art and Nature School.

Christine Wu is a Chinese-Canadian poet whose work has appeared in a variety of literary periodicals.

Lightning Source UK Ltd.
Milton Keynes UK
UKHW020926091221
395320UK00003B/56